T0281154

Studies in Computational Intelligence

Volume 742

Series editor

Janusz Kacprzyk, Polish Academy of Sciences, Warsaw, Poland
e-mail: kacprzyk@ibspan.waw.pl

The series "Studies in Computational Intelligence" (SCI) publishes new developments and advances in the various areas of computational intelligence—quickly and with a high quality. The intent is to cover the theory, applications, and design methods of computational intelligence, as embedded in the fields of engineering, computer science, physics and life sciences, as well as the methodologies behind them. The series contains monographs, lecture notes and edited volumes in computational intelligence spanning the areas of neural networks, connectionist systems, genetic algorithms, evolutionary computation, artificial intelligence, cellular automata, self-organizing systems, soft computing, fuzzy systems, and hybrid intelligent systems. Of particular value to both the contributors and the readership are the short publication timeframe and the world-wide distribution, which enable both wide and rapid dissemination of research output.

More information about this series at http://www.springer.com/series/7092

Tokuro Matsuo · Tsunenori Mine
Sachio Hirokawa

Editors

New Trends in E-service and Smart Computing

 Springer

Editors
Tokuro Matsuo
Advanced Institute of Industrial Technology
Tokyo
Japan

Sachio Hirokawa
Kyushu University
Fukuoka
Japan

Tsunenori Mine
Kyushu University
Fukuoka
Japan

ISSN 1860-949X   ISSN 1860-9503   (electronic)
Studies in Computational Intelligence
ISBN 978-3-319-88971-9   ISBN 978-3-319-70636-8   (eBook)
https://doi.org/10.1007/978-3-319-70636-8

Printed on acid-free paper

This Springer imprint is published by Springer Nature
The registered company is Springer International Publishing AG
The registered company address is: Gewerbestrasse 11, 6330 Cham, Switzerland

Preface

This book includes theory and practice on emerging technologies in e-service and artificial intelligence from an academic and professional viewpoint. In Chapter ""Co-occurrence Relation" and "Ingredient Category" Recommend Alternative-Ingredients," authors study on contents recommendation on cooking where several ingredients are used. When recipes are defined and some ingredients are not prepared, the system proposes alternative ingredients using data on co-occurrence relation and ingredients categories. The result of evaluation shows high effectiveness and tried to use the recommended ingredients in actual cooking workshop. Chapter "Design and Initial Evaluation of Bluetooth Low Energy Separate Channel Fingerprinting" presents a study on location information inference using data acquired by low-energy Bluetooth devices. The proposed algorithm employs channel-specific features in fingerprinting and is evaluated in the condition where sensor devices are positioned in a room. Chapter "Toward Sustainable Smart Mobility Information Infrastructure Platform: Project Overview" introduces the sustainable information infrastructure project for smart mobility systems. The project contains two features on applying lifecycle-oriented system development methods to real world and dealing with uncertainty in system design and development upper phase. Chapter "Model-Based Methodology Establishing Traceability Between Requirements, Design and Operation Information in Lifecycle-Oriented Architecture" presents a lifecycle-oriented development process which improves the requirement and design in terms of uncertainties for realizing sustainable information architecture for smart mobility. Authors show a case study using the proposed development process to a transfer guide app Patrash. Chapter "Sports Game Summarization Based on Sub-events and Game-Changing Phrases" introduces a summarization task of sports events on Twitter, focusing on an abstractive approach based on sub-events in the sports event. An experiment suing the proposed method shows the effectiveness of the sports game summarization method as compared with related work based on an extractive approach. Chapter "Headline Generation with Recurrent Neural Network" is aimed at generating a headline using a recurrent neural network which is based on a machine translation approach. Experiments show the effectiveness of the proposed method, which can generate appropriate headlines but in some articles this method generate meaningless headlines. Chapter "Customer State Analysis with Enthusiasm Analysis" introduces a customer behavior analysis using enthusiasm analysis, which estimates customers' activation levels. Author estimates enthusiasm levels, which denote customers' activation, from observations and applies them to prediction of discovery of drop-off users. The result of evaluation confirms many drop-off users took lower enthusiasm levels in evaluation point and the enthusiasm level could be used to predict drop-off users.

Dr. Matsuo, Dr. Tsunenori Mine, and Dr. Sachio Hirokawa are grateful to the authors and reviewers for their contribution to this work. Editors also acknowledge with their gratitude the editorial team of Springer-Verlag for their support during the preparation of the manuscript.

Tokyo, Japan   Dr. Tokuro Matsuo
Fukuoka, Japan   Dr. Tsunenori Mine
Fukuoka, Japan   Dr. Sachio Hirokawa
July 2017

Contents

"Co-occurrence Relation" and "Ingredient Category" Recommend
Alternative-Ingredients   1
Naoki Shino, Ryosuke Yamanishi and Junichi Fukumoto

Design and Initial Evaluation of Bluetooth Low Energy Separate
Channel Fingerprinting   19
Shigemi Ishida, Yoko Takashima, Shigeaki Tagashira and Akira Fukuda

Toward Sustainable Smart Mobility Information Infrastructure
Platform: Project Overview   35
Akira Fukuda, Kenji Hisazumi, Tsunenori Mine, Shigemi Ishida,
Takahiro Ando, Shota Ishibashi, Shigeaki Tagashira, Kunihiko Kaneko,
Yutaka Arakawa, Weiqiang Kong and Guoqiang Li

Model-Based Methodology Establishing Traceability Between
Requirements, Design and Operation Information in
Lifecycle-Oriented Architecture   47
Shota Ishibashi, Kenji Hisazumi, Tsuneo Nakanishi and Akira Fukuda

Sports Game Summarization Based on Sub-events
and Game-Changing Phrases   65
Yuuki Tagawa and Kazutaka Shimada

Headline Generation with Recurrent Neural Network   81
Yuko Hayashi and Hidekazu Yanagimoto

Customer State Analysis with Enthusiasm Analysis   97
Hidekazu Yanagimoto

"Co-occurrence Relation" and "Ingredient Category" Recommend Alternative-Ingredients

Naoki Shino, Ryosuke Yamanishi and Junichi Fukumoto

Abstract Websites and magazines are now becoming more useful for general people to cook a dish everyday. Sometimes, we have inconvenience for such information because the listed ingredients in the recipe can not be prepared. This paper proposes a recommendation method of alternative ingredients towards such situation. The recommendation is realized by considering co-occurrence of ingredients on recipe database and ingredient category stored in a cooking ontology. The subjective evaluation experiments showed 88% appropriateness for the alternative-ingredients recommended by our proposed method. Also, the recommended ingredients were used in a real cooking, the good tastes were confirmed through the workshop.

1 Introduction

Cooking dishes is a kind of creative activity in which both ingredients and methods are selected and combined. The complex combination sometimes provides us with deep emotion and reminds us of the home country: the taste of mother and traditional taste. Cooking recipes have been stored to make a good tasty dish again and again, but it used to be difficult to get such recipes for good taste dishes. However, getting a good recipe would become easier because of some recipe website such as "Cookpad[1]" and "allrecipes.com.[2]" The number of users checking recipes every day recently increases; the number Cookpad users is about 56,440,000 on 2015.

---

[1] http://cookpad.com/.
[2] http://allrecipes.com/.

N. Shino · R. Yamanishi (✉) · J. Fukumoto
Ritsumeikan University Information Science and Engineering, Nojihigashi 1–1–1,
Kusatsu, Shiga, Japan
e-mail: ryama@media.ritsumei.ac.jp

N. Shino
e-mail: n_shino@nlp.is.ritsumei.ac.jp

J. Fukumoto
e-mail: fukumoto@media.ritsumei.ac.jp

© Springer International Publishing AG 2018
T. Matsuo et al. (eds.), *New Trends in E-service and Smart Computing*,
Studies in Computational Intelligence 742, https://doi.org/10.1007/978-3-319-70636-8_1

1

Recipes on Website generally show both the ingredients that should be prepared for a dish and the procedure of the cooking. Some of the listed ingredients, sometimes, can not be used for the cooking because of some reasons: dislikes, allergies, religions, decreasing in the catch, and not sold in the neighbor. It is difficult for users to search recipes the users like while preventing to use such ingredients. Some of them alternatively use another ingredient instead of a non-available ingredient; we define such an ingredient as an *alternative-ingredient* in this paper. To think up appropriate alternative-ingredients needs much knowledge and experience for ingredients. For even experts for cooking, it should be difficult to choose alternative-ingredients considering a complex relationship between ingredients in a recipe.

The goal of this study is to enhance both beginners and experts for cooking with convenience and creativity provided by using alternative-ingredient. In this paper, we propose a recommendation method for alternative-ingredients. The proposed method is a mix of machine learning and knowledge engineering techniques. The varied expressions of ingredients would be merged according to the proposed cleansing rule with an ingredient ontology. The relationships among ingredients are acquired from co-occurrence relations in a recipe database.

2 Related Work

Recipe recommendation has been reported in some existing papers. Ueta et al. has developed a recipe recommendation system based on nutrition information [1]. Their system does not only recommend recipes but also compose a database of ingredients nutrition. The recipes recommended by their system contains effective nutrition for the user's requirement such as bad point of his/her body and a kind of symptoms. Then a database of ingredients nutrition is structured while using their recommendation system. Freyne et al. propose a recipe recommendation system considering the preference of users for diet [2]. Their system is based on collaborative filtering. Sobecki et al. propose a recipe recommendation system which uses demographic, content-based, and collaborative filtering [3]. Above studies are for the recommendation of recipe against the goal of this study which is the recommendation of alternative-ingredient. This paper focuses on the inside of a recipe, and try to help users to arrange an existing recipe by themselves using alternative-ingredient.

There are some methods to find an alternative-ingredient. Cojan et al. developed the project called "TAABLE3" in "Computer Cooking Contest." In TAABLE3, they constructed a search system for alternative ingredients based on information of alternative ingredients on web [4]. Teng et al. structured substitute ingredients networks based on usage on ingredients on recipe database [5]. This method discovers alternative ingredients from several recipes of each dish. Though both suitability with other ingredients and similarity with the exchanged-ingredient are important in choosing alternative-ingredient, such methods do not take them in their consideration.

Chef Watoson [6] recommends additional ingredients for a dish based on analysis of recipe database. The additional ingredients are recommended by Chef Watson is based on the research about the flavor of ingredients by Ahn et al. [7]. However, the recommended ingredients might be effective for only western cuisines. In eastern cuisines, easter spices used in dishes are very strong and the flavors are not so much focused on.

3　Basic Idea

3.1　Two Types of Criteria for Alternative-Ingredients

This paper prepares the following two types of criteria whether a food can be used as an alternative-ingredient or not.

- The similarity with the exchange-ingredient.
- Compatibility with ingredients used together in the recipe.

Figure 1 shows the idea of an alternative-ingredient based on these criteria.

Cooking experts often use ingredients similar to an exchange-ingredient as the alternative-ingredient; for example, chicken breast is used instead of shrimp, and avocado is used instead of raw tuna. The texture of such alternative-ingredient is similar to the one of exchange-ingredient. Cooking experts use these ingredients to cook a dish for making the same taste on the recipes.

Alternative-ingredients are also used to cook a new delicious dish, which is different from an original recipe; for example, the eringi is used instead of the onion of the hamburger. In this case, though the eringi is far from the onion, eringi is used as the alternative-ingredients based on the compatibility with minced meats. This

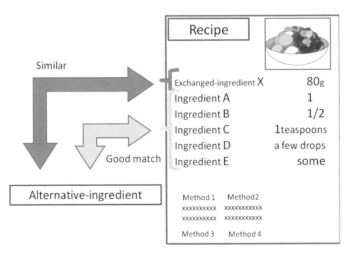

Fig. 1 The general idea of the criteria for alternative-ingredients. Appropriate alternative-ingredients should satisfy the both or either condition: "similarity with exchange-ingredient" and "compatibility with ingredients used together in the recipe"

paper proposes a method considering similarity and compatibility to recommend various alternative-ingredients. We believe that the two types of criteria are effective for recommending alternative-ingredients.

3.2   Similarity Based on Ingredients Category

Ingredients have several characteristics: taste, texture, flavor, nutrition, and color. Therefore it is difficult to find out ingredients similar to a given ingredient. Iwatani et al. propose a method for food texture evaluation using an accelerometer sensor. But, this method is not considered sensibility of the individual of users.

This paper uses five ingredient categories in an ontology [8]: meats, vegetables, fishes, seasoning, and the others. And, this paper defines ingredients in the same category as similar ingredients with each other. The definition is based on the idea that texture and nutrition of ingredients in the same category are more similar than the ones of the ingredients in other categories.

3.3   Compatibility Based on Co-occurrence Relation Among Ingredients

This paper defines that relative frequency that two ingredients are concurrently used in a recipe (i.e., co-occurrence relation). In this paper, the combination of ingredients with high co-occurrence frequency are regarded as the compatible combination. For example, onion and bacon should be a compatible combination, and the combination is used in various dishes: cream stew, German potato, spaghetti and others. Whether an ingredient can be the alternative-ingredient or not is determined by referring co-occurrence of the two ingredients and others in a recipe which can be calculated from a recipe database.

4   The Proposed System

This paper proposes a recommendation system for alternative-ingredient using co-occurrence frequency and ingredient categories. The system consists the following four steps.

1. Calculating occurrence frequency of a single ingredient.
2. Calculating co-occurrence frequency of a combination of two ingredients.
3. Verifying whether the category of the focused ingredient is same to the one of the exchange-ingredient or not.
4. Filtering by co-occurred seasoning.

Table 1 Example of ingredients table of the cookpad_data

Recipe ID	Ingredients name	Quantity
ad7d585b06850f8823ff21bb1	Ginger	Some
ad7d585b06850f8823ff21bb1	Garlic	Some
ad7d585b06850f8823ff21bb1	Green onion	1 stick
ad7d585b06850f8823ff21bb1	Pork	2 block
ad7d585b06850f8823ff21bb1	Sugar	1~2 teaspoons
ad7d585b06850f8823ff21bb1	Cooking wine	Lots
ad7d585b06850f8823ff21bb1	Soy sause	Suitable
4afce5687dc18651cd06e264	Carrot	2 sticks
4afce5687dc18651cd06e264	Green pepper	4
4afce5687dc18651cd06e264	Lotus root	1
4afce5687dc18651cd06e264	Minced pork	250 g
4afce5687dc18651cd06e264	Pepper	A few drops
4afce5687dc18651cd06e264	Salt	A few drops

This paper uses ingredients description in recipe provided by Cookpad, i.e., cookpad_data. The cookpad_data has 1,715,595 Japanese recipes.

4.1 Calculating Occurrence Frequency of One of the Ingredients

It is necessary to count the occurrence of each ingredient to calculate co-occurrence frequency. To count the occurrence of each ingredient, "ingredients" table of cookpad_data is used. Table 1 shows a sample of "ingredients" table. In "ingredients" table, there are three types of information: name of ingredients, ID of the recipe where the ingredient is used, a quantity of an ingredient in the recipe. In the table, same ingredients appear multiple times because this table was made from each ingredient of each recipe, it is enabled to search a recipe from an ingredient. Using recipe ID, combinations of two ingredients on the same recipe can be listed. From the list, the frequency of each ingredient can be obtained.

4.2 Solving Orthographical Variant with a Cooking Ontology

An orthographical variant would be a problem, especially in Japanese. This paper solves the orthographical variant by cooking ontology. A cooking ontology has the hierarchical structure of ingredients. Table 2 shows a sample of cooking ontology.

Table 2 Example of ingredients data on cooking ontology

Category	Broad ingredient name	Narrow ingredient name
Vegetables	ゴボウ (Burdock)	ゴボウ (Burdock in Katakana)
Vegetables	ゴボウ	ごぼう (Burdock in Hiragana)
Vegetables	ゴボウ	牛蒡 (Burdock in Kanji)
Vegetables	ゴボウ	新ゴボウ (New Burdock)
Vegetables	ゴボウ	新ごぼう (New Burdock)
Vegetables	ゴボウ	新牛蒡 (New Burdock in Kanji)
Vegetables	れんこん (Lotus root)	れんこん (Lotus root in Hiragana)
Vegetables	れんこん	レンコン (Lotus root in Katakana)
Vegetables	れんこん	蓮根 (Lotus root in Kanji)
Sea food	しじみ (Freshwater clam)	しじみ (Freshwater clam in Hiragana)
Sea food	しじみ	シジミ (Freshwater clam in Katakana)
Sea food	しじみ	蜆 (Freshwater clam in Kanji)

The ontology can be divided into three types of the level; ingredient category, broad ingredient name, and narrow ingredient name. The orthographical variant is solved to standardize narrow ingredient name to broad ingredient name.

The cooking ontology covers 84% (37,580/44,734) number of ingredients of Cookpad recipe by perfect matching. This paper proposes a method using a partial match to cover more ingredients. The rules of the method are as follows;

At first, the method checks whether an ingredient partially matched with only one ingredient on an ontology or not.

1. Return the broad ingredient if it is partially matched with only one ingredient on an ontology.

Otherwise, i.e., the ingredients is partially matched with multiple ingredients on an ontology, the method checks whether an ingredient is included in the end part of the ingredient or not.

2. Return the broad ingredient of the end part of an ingredient if the ingredient has an ingredient at the end part.
3. Return the broad ingredient of the firstly occurred ingredient if the ingredient does not have an ingredient at the end part.

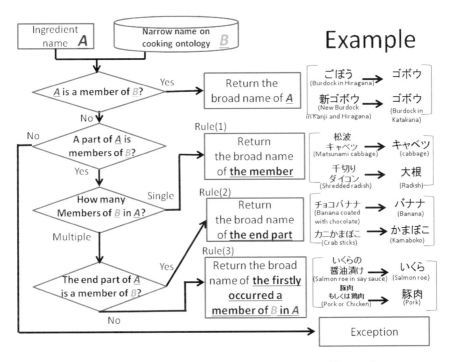

Fig. 2 The image of a name association method for ingredient with cooking ontology

Figure 2 shows the idea of this method and example of the ingredients covered by each rule. The ingredients covered by the rule (1) are ingredient with additional information: locality e.g. "松波キャベツ (cabbage harvested in Matsunami)," cooking method e.g. "千切りキャベツ (shredded cabbage)," how to use e.g. "サラダ用キャベツ (cabbage for salad)," symbol e.g. "★キャベツ (★cabbage)." The ingredients covered by the rule (2) are ingredients that mimicking other food e.g. "カニかまぼこ (crab sticks)" and "タコさんウインナー (sausage cut in the shape of an octopus)," and ingredients seasoned by other food e.g. "塩数の子 (Salted herring roe)" and "チョコバナナ (banana coated with chocolate)." The main of such ingredient is described in the end part. The ingredients covered by the rule (3) are ingredients that ingredients seasoned by other food with the cooking method e.g. "いくらの醤油漬け (salmon roe in soy sauce)," and ingredients with additional information of the alternative-ingredient e.g. "鶏肉または豚肉 (chicken or pork)." In this case, the firstly occurred ingredient is the main, and the others are the sub.

This method increases the number of the standardized ingredient from 84.0 to 95.3%. Here, most of the missed ingredients are minor ingredients in Japan such as "endive," ingredients categories such as "meat," or no longer ingredients such as "memories."

5 Experiment to Recommend Alternative-Ingredients

5.1 Co-occurrence Frequency of Combination of Two Ingredients

The most recipe is structured of multiple ingredients. Therefore, it is necessary to consider co-occurrence frequency of multiple combinations for choosing compatible ingredients for the recipe. This paper applies Naive Bayes classifier to calculate the co-occurrence frequency. Naive Bayes classifier [9] is effective for a problem that multiple characteristics affect to each other [10]. This paper assumes "choosing the ingredient often co-occurred with other ingredients in a recipe" as "classification of the ingredient into recipes using the other ingredients." Then, each ingredient has the likelihood that the ingredient belongs to each recipe.

For a good ingredient recommendation, compatibility score between the ingredient and R_i (i.e. ith recipe) is given to all of K kinds of ingredients, which occurred in the recipe database. The Eq. (1) calculates the compatibility S_{k,R_i} with $N_{F_{i,n}}$ concerning the frequency of a single ingredient $F_{i,n}$ and N_k concerning the frequency of the focused ingredient k;

$$S_{k,R_i} = (\prod_{n=0}^{r} \frac{CoOc(k, F_{i,n})}{N_{F_{i,n}}}) * N_k, \tag{1}$$

where, $CoOc(k, F_{i,n})$ shows the co-occurrence frequency of ingredient k with each ingredient $F_{i,0} \sim F_{i,r}$ in recipe R_i. $CoOc(k, F_{i,n})$ can be calculated as the following equation;

$$\frac{CoOc(k, F_{i,n})}{N_{F_{i,n}}} = \begin{cases} \frac{CoOc(k, F_{i,n})}{N_{F_{i,n}}} & (N_{F_{i,n}} > 0), \\ 1 & (N_{F_{i,n}} = 0), \end{cases} \tag{2}$$

where, $N_{F_{i,n}} = 0$ means that an ingredient $F_{i,n}$ is not in the recipe database.

If an ingredient used in a recipe is not co-occurred with the focused ingredient x in a recipe database, the S_{k,R_i} would be 0 caused by $CoOc(k, F_{i,n}) = 0$. In such case, this paper thus makes $CoOc(k, F_{i,n}) = 0.0001$ as the infinitesimal score as the following equation;

$$CoOc(k, F_{i,n}) = \begin{cases} CoOc(k, F_{i,n}) & (CoOc(k, F_{i,n}) > 0), \\ 0.0001 & (CoOc(k, F_{i,n}) = 0). \end{cases} \tag{3}$$

The proposed system recommends alternative-ingredients based on $GM_{x \in k, R_i}$, which concerns the proportion of $S_{x \in k, R_i}$ with compatibility score of all ingredients in a recipe database. This idea comes from that the proportion can be assumed as the likelihood that an ingredient x occurs with ingredients $F_{i,n}$ in recipe R_i. The $GM_{x \in k, R_i}$ is calculated as follows;

$$GM_{x\in k,R_i} = \frac{S_{x\in k,R_i}}{\sum_{k=0}^{K} S_{k,R_i}}. \tag{4}$$

5.2 Improvement of the Recommendation with Semantics of Ingredient

The scoring described in Sect. 5.1 lists candidates of the alternative-ingredient with a lack of coherences. Therefore, the ingredients belonging to a same category with the exchanged-ingredient are extracted from the candidates with a cooking ontology. The cooking ontology covers five types of ingredient category: meat, vegetable, seafood, seasoning and the others.

Seasoning should be a key of a dish. If a dish includes a miss-matched combination of an ingredient and a seasoning, the dish would be usually felt bad-taste by most of the people. The proposed system excepts the combination of ingredient and seasoning that does not co-occur in a recipe database.

In order to verify the effectiveness of the proposed system, the evaluation experiments were conducted. The experiments had two points for discussion: whether output ingredient is suitable as alternative-ingredient or not, and the effectiveness of recommending the ingredients of the same category with the exchange-ingredient. The recommendations of the alternative-ingredient were conducted for four recipes. The top 10 ingredients for each category (meat, fish, and vegetable) recommended by the proposed system would be discussed in this paper; all ingredients would be discussed if less than 10 ingredients were recommended for each category. Tables 3, 4, 5 and 6 show the recommended alternative-ingredients, their ranks and scores.

5.3 Subjective Evaluations

In the experiment, each three males and females participated. Table 7 shows the questionnaires for the participants. All of them were their twenties; they cook one or more times per week. Alternative-ingredients and the information of recipe for each alternative-ingredient (i.e. pictures of the dish, the list of the ingredients, and the exchanged-ingredient) were presented to the participants. The participants evaluated whether dish using the recommended alternative-ingredient looked "absolutely good," "maybe good," "maybe bad," or "absolutely bad." And, the participants evaluated whether the recommended ingredients were suitable as "alternative" or not based on their feeling. The alternative-ingredients were presented in different order for each participant.

The subjective evaluations were discussed with the statistics. Tables 8, 9, 10, and 11 each shows the results of the subjective evaluation for alternative-ingredients of same category of the exchanged-ingredients for each recipe, respectively. In the table,

Table 3 Alternative-ingredients for pork in cream stew

Rank	Meat		Sea food		Vegetable	
	Ingredients name	Opponent	Ingredients name	Opponent	Ingredients name	Opponent
1	Chicken	36.9	Shrimp	91.3	Garlic	97.3
2	Ham	32.7	Salmon	8.7	Ginger	1.4
3	Minced meat	30.5	Cod	0.0	Tomato	0.8
4	Beef	0.0	Tuna	0.0	Green onion	0.5
5	Sausage	0.0	Scallop	0.0	Green pepper	0.1
6			Red snapper	0.0	Crab	0.0
7			Small clam	0.0	Corn	0.0
8			Cod	0.0	Shiitake mushroom	0.0
9			Octopus	0.0	Red pepper	0.0
10			Big clam	0.0	Cabbage	0.0

Table 4 Alternative-ingredients for Bamboo shoot in Chinjaorosu

Rank	Meat		Sea food		Vegetable	
	Ingredients name	Opponent	Ingredients name	Opponent	Ingredients name	Opponent
1	Chicken	86.6	Shrimp	98.7	Onion	31.2
2	Minced meat	11.6	Squid	1.1	Garlic	26.6
3	Ham	1.6	Cod	0.0	Ginger	22.1
4	Beef	0.2	Swordfish	0.0	Green onion	13.2
5	Sausage	0.0			Carrot	5.0
6					Shiitake mushroom	0.7
7					Tomato	0.7
8					Eggplant	0.2
9					Cabbage	0.2
10					Red pepper	0.1

Table 5 Alternative-ingredients for Grape tomato in Green salad

Rank	Meat		Sea food		Vegetable	
	Ingredients name	Opponent	Ingredients name	Opponent	Ingredients name	Opponent
1	Chicken	99.0	Tuna	94.3	Tomato	70.0
2	Beef	0.6	Shrimp	5.6	Garlic	21.1
3	Pork	0.5	Scallop	0.0	Green onion	3.8
4			Shirasu	0.0	Ginger	2.5
5			Salmon	0.0	Carrot	1.0
6			Squid	0.0	Potato	0.8
7			Small clam	0.0	Green pepper	0.5
8			Mackerel	0.0	Perilla	0.1
9			Garfish	0.0	Radish	0.1
10					Cabbage	0.0

Table 6 Alternative-ingredients for Carrot in Kakiage Tempura

Rank	Meat		Sea food		Vegetable	
	Ingredients name	Opponent	Ingredients name	Opponent	Ingredients name	Opponent
1	Chicken	83.9	Squid	94.2	Green onion	58.5
2	Minced meat	12.2	Scallop	4.2	Ginger	29.4
3	Pork	3.9	Salmon	1.3	Carrot	11.5
4	Ham	0.0	Shirasu	0.3	Siitake mushroom	0.5
5	Beef	0.0	Tuna	0.0	Garlic	0.1
6	Sausage	0.0	Cod	0.0	Perilla	0.0
7			Crab	0.0	Radish	0.0
8			Small cram	0.0	Tomato	0.0
9			Octopus	0.0	Red pepper	0.0
10			Red snapper	0.0	Green pepper	0.0

"Taste-score" shows the average of evaluation for the dish looks good taste; in this paper the score was defined as "absolutely good = 0," "maybe good = 2," "maybe bad = 1," and "absolutely bad = 0." "Alternativeness-score" shows the average of the evaluations for suitability as the alternative-ingredient; "suitable = 1" and "not suitable = 0."

Table 12 shows the percentage of ingredients showing more than 1.5 and less than 1.5 taste-score; here 1.5 is a median of 0 and 3. If the taste score is more than 1.5,

Table 7 The list of questionnaire items

Item	Answer content
Gender	Male or female
Frequency of cooking	How many days between cookings
Deliciousness of the dish with alternative-ingredients	(A) absolutely good
	(B) maybe good
	(C) maybe bad
	(D) absolutely bad
Suitability for alternative ingredients	Suitable or not-suitable

Table 8 Evaluation for the alternative-ingredients for pork in cream stew

Rank	Ingredients name	Suitability	Deliciousness
1	Chicken	1.00	3.00
2	Ham	0.83	2.83
3	Minced meat	1.00	2.83
4	Beef	1.00	2.50
5	Sausage	1.00	3.00

Table 9 Evaluation for the alternative-ingredients for Bamboo shoot in Chinjaorosu

Rank	Ingredients name	Suitability	Deliciousness
1	Onion	1.00	2.83
2	Garlic	0.50	2.00
3	Ginger	0.33	2.50
4	Green onion	1.00	2.83
5	Carrot	1.00	2.67
6	Shiitake mushroom	0.83	1.67
7	Tomato	0.17	1.33
8	Eggplant	0.67	2.33
9	Cabbage	0.83	2.5
10	Red pepper	0.50	2.17

the ingredient was regarded as looking good. As the result, 88% of ingredients was evaluated as looking good.

Table 13 shows the average of alternativeness-score for each category. The scores that the category of the exchanged-ingredient and alternative-ingredient are equaled were higher than the others. It was suggested that the ingredients belonging to the same category with exchange-ingredient were more suitable as alternativeness than the others.

Table 10 Evaluation for alternative-ingredients for Grape tomato in Green salad

Rank	Ingredients name	Suitability	Deliciousness
1	Garlic	1.33	1.33
2	Green onion	0.50	2.17
3	Ginger	0.33	2.00
4	Carrot	0.83	2.83
5	Potato	0.67	2.17
6	Green pepper	0.50	2.33
7	Perilla	0.67	2.50
8	Radish	0.83	2.67
9	Cabbage	0.50	2.17

Table 11 Evaluation for alternative-ingredients for Carrot in Kakiage tempura

Rank	Ingredients name	Suitability for alternative-ingredients	Deliciousness
1	Green onion	1.00	2.67
2	Ginger	0.83	2.67
3	Shiitake mushroom	0.83	2.33
4	Garlic	0.50	1.33
5	Perilla	0.83	2.83
6	Radish	0.67	1.67
7	Tomato	0.33	1.00
8	Red pepper	0.33	1.83
9	Green pepper	0.67	1.83

Table 12 The percentage of the evaluation of alternative-ingredients

Alternative-ingredients evaluated as good	29 (88%)
Alternative-ingredients evaluated as bad	4 (12%)

Table 13 Evaluation for suitability of alternative-ingredients

Dishes	Cream stew	Chinjaorosu	Green salad	Kakiage tenpura
Exchanged-ingredients	Pork	Bamboo shoot	Grape tomato	Carrot
Category	Meat	Vegetable	Vegetable	Vegetable
Meat	*0.97*	0.10	0.72	0.31
Sea food	0.75	0.25	0.56	0.5
Vegetable	0.27	*0.68*	*0.62*	*0.67*

5.4 Discussions for the Characteristics of Ingredients

The taste score of "grape tomato" in the green salad was low. The one of the reason was considered that the taste of "grape tomato" was stronger than other vegetables. The feelings for the salad taste might become week by using weak taste vegetable, and it would make the lower evaluation. As shown in Table 13, meat was better than vegetable as alternative-ingredients for grape tomato. From these considerations, it was suggested that not only ingredient category but also the strength of taste were taken in the consideration for alternative ingredients.

The evaluation of garlic, ginger, green onion, red pepper, and others were low in Tables 8, 9, 10, and 11. These ingredients occurred in more recipes than other ingredients because these are also used as spices. The system should adjust the recommendation scores for such particular ingredients.

Through the experiments, some participants commented "good-match ingredients is less because the ingredients suitable with pumpkin is less" for stew. From this comment, it was suggested that the importance of each ingredient should be taken in the consideration; what is the main ingredient of a dish should be considered.

6 Workshop

We conducted the workshop where the participants actually taste the dishes using alternative-ingredients recommended by the proposed method. The participants had dishes and answered to questionaries. The dishes using the recommended alternative-ingredients were cooked by one of the authors.

6.1 Workshop Environments

In the workshop, black olive is set as an exchanged ingredient in a recipe for potato salad stored in COOKPAD. Figure 3 shows the snapshot and the ingredients of the recipe. Potato salad consists of many ingredients. Some ingredients would be into our mouth together, and the original textures of ingredient remained in potato salad without melting and being mixed. These mean that the recommended alternative-ingredient can be individually evaluated. For these reasons, we selected and used the potato salad recipe in a workshop.

Black olive in this recipe is an ingredient that is hardly used in Japanese native dishes. In this experiment, black olive is assumed as an exchange-ingredient. This workshop prepared a dish cooking just according to the recipe (hereafter, original dish) and dishes cooking with the recommended alternative-ingredients instead of black olive (hereafter, alternative dish). Alternative-ingredients used in alternative dishes were tomato, eggplant, chestnut, cod, and minced meat. Table 14 shows the

Fig. 3 The snapshot of the recipe used for the workshop and its ingredients

Little spicy potato salad with olibe for adult

Ingredients(for 2 or 3 people)

2 potatos
5 or 6 black olive
30 to 40 gram sliced onion
1 bacon
1 tbsp mayonnaise
1 tsp lemon juice
1 tsp ponzu sauce
1 tsp sweet chili sauce
to taste flesh parsley
to taste tabasco
to taste cresol
to taste garlic powder
to taste black pepper

Table 14 Alternative ingredients recommended by the proposed method

Rank	Meat		Sea food		Vegetable	
	Ingredients name	Opponent	Ingredients name	Opponent	Ingredients name	Opponent
1	Minced meat	68.3	Shrimp	92.8	Tomato	91.6
2	Pork	29.9	Cod	2.6	Carrot	6.7
3	Chicken	1.8	Salmon	2.4	Green onion	1.0
4	Sausage	0.0	Clams	1.6	Green pepper	0.6
5	Beaf	0.0	Tuna	0.2	Mushroom	0.0
6			Scallops	0.1	Eggplant	0.0
7			Shine	0.1	Letas	0.0
8			Octopus	0.1	Lemon	0.0
9			Squid	0.0	Red pepper	0.0
10			Sardine	0.0	Pumpkin	0.0

outputs; Chestnut was in Rank 11 of vegetables. From the output of the proposed method, we selected the ingredients that might be effective and the ingredients that people generally do not think up. Some ingredients can not be eaten raw, and some of the others are too big to eat. For such ingredients, we have to apply cooking methods. As shown in the Table 15, one of the authors subjectively chose cooking methods for alternative-ingredients: cutting, boiling, baking, and the others.

Ten to seventy years old people had alternative dish and original dish for evaluation. They compared the tastes and evaluated which dish is better. As the first step, the participant had original dish. Next, they chose one of five alternative dishes and

Table 15 Cooking methods for alternative ingredients on this workshop

Alternative ingredients	Cooking methods
Tomato	Mashed
Eggplant	Cut and grilled
Chestnut	Cut and baked
Cod	Cut and boiled
Minced meat	Fried

Table 16 Evaluation for the alternative-ingredients for Bamboo shoot in Chinjaorosu

Alternative ingredients	Worse	Equal	Better
Tomato	41.7% (5)	8.3% (1)	50.0% (6)
Eggplant	20.0% (2)	40.0% (4)	40.0% (4)
Chestnut	0.0% (0)	18.2% (2)	81.8% (9)
Cod	27.3% (3)	36.7% (4)	36.7% (4)
Minced meat	8.3% (1)	16.7% (2)	75.0% (9)

had it. At the last, they evaluated whether the alternative dish is better than original dish or not: the options are "better," "equal" and "worse."

6.2 Results

Fifty six participants cooperated in the workshop. Table 16 shows the results. The table shows the probability of each evaluation level for each food; number in round brackets shows the actual number of the evaluations.

We believe that dishes which is better or equals to the original dish would be required for the alternative dish. In this workshop, "better" and "equal" are assumed as good evaluation. The result shows that 80.4% (45 of 56) were good evaluation in the workshop.

6.3 Discussions

Most alternative dishes were evaluated as "better" or "equal." Especially, the dish using chestnut were evaluated as "better" or "equal" by all participants. Chestnut might have a texture similar to a potato and often used in so many kinds of dishes: main dishes, side dishes, and desserts. Those seemed to be a reason why chestnut can obtain good evaluations. The ingredient evaluated as the second best was minced meat. It is well known that the combination of minced meat, potato, and onion is used in croquettes. So, this dish might be accepted by many participants without any discomforts.

Evaluations of cod and eggplant were even; these ingredients have weak taste by itself. It seemed that the combination of potato and these ingredients was almost bad beating in a potate salad though the ingredients had good relations to other ingredients in the recipe.

The dish using tomato obtained many "better" and "worse." For the "worse" evaluation, the comments for the dish using tomato were follows; *"Tomato is too strong"* and *"This is not potato salad."* Against, *"This is good because of freshness"* was commented as the "better" evaluation. This differences of evaluations let us find that evaluation greatly depended on participants each preference for alternative-ingredient. How to evaluate the ingredients considering preferences will be our future work.

Evaluation for minced meat was better than average evaluation of vegetable which was the same category of the exchanged-ingredient: black olive. From this results, it was suggested that category of the exchange-ingredient is not always essential.

7 Conclusions

This paper proposed alternative-ingredients recommendation based on co-occurrence frequency of ingredients and cooking ontology. The experiments suggested two findings: (1) the ingredients often co-occurred with other ingredients in a recipe were good as the altenative-ingredients, and (2) the ingredients belonging to the same category of the exchanged-ingredient were more suitable as the alternative-ingredient. Through the workshop, the effectiveness of most alternative-ingredients was confirmed with actual tasting. And, it was found that the individual preference for ingredients should be taken in the recommendation mechanism. According to the results and opinions, two points to be improved were found: (1) strength of taste and (2) the main ingredient of a dish should be taken into the consideration. In our future, ingredients roll and texture of ingredients described in recipe review would be applied into the recommendation.

Acknowledgements Yamanishi's work was supported in part by Artificial Intelligence Research Promotion Foundation. And, in this paper, we used recipe data provided by Cookpad and the National Institute of Informatics.

References

1. Ueta, T., Iwakami, M., Ito, T.: A recipe recommendation system based on automatic nutrition information extraction. In: International Conference on Knowledge Science, Engineering and Management, pp. 79–90 (2011)

2. Freye, J., Berkovsky, S.: Recommendation food: reasoning on recipes and ingredients. In: Proceedings of the 18th International Conference on User Modeling, Adaption, and Personalization, pp. 381–386 (2010)
3. Freye, J., Shlomo, B.: Intelligent food planning: personalized recipe recommendation. In: Proceedings of the 2010 International Conference on Intelligent User Interfaces, pp. 321–324 (2010)
4. Blansche, A., Cojan, J., Dufour-Lussier, V., Lieber, J., Molli, P., Nauer, E., Skaf-Molli, H., Toussaint, Y.: Taaable 3: adaptation of ingredient quantities and of textual preparations. In: 18th International Conference on Case-Based Reasoning - ICCBR 2010 Computer Cooking Contest Workshop Proceedings, pp. 189–198 (2010)
5. Teng, C.-Y., Lin, Y.-R., Adamic, L.A.: Recipe recommendation using ingredient networks. In: Proceedings of the 4th Annual ACM Web Science Conference, pp. 298–307 (2012)
6. IBM: Cognitive cooking. http://www.ibm.com/smarterplanet/us/en/cognitivecooking/ (2014)
7. Ahn, Y.-Y., Ahnert, S.E., Bagrow, J.P., Barabasi, A.-L.: Flavor network and the principles of food paring. Sci. Rep. **1**(196), 1–7 (2011)
8. Yamakata, Y., Imahori, S., Sasada, T., Mori, S.: Motion name ontology by structural mapping of processing on recipes with a same titles. In: HCG Symporsium 2014, pp. C–8–5 (8 p) (2014). (In Japanese. The title is translated by the authors of this paper)
9. Maron, M.E.: Automatic indexing: an experimental inquiry. J. ACM (JACM) 404–417 (1961)
10. Domingos, P., Pazzani, M.: On the optimality of the simple bayesian classifier under zero-one loss. Mach. Learn. 103–130 (1997)

Design and Initial Evaluation of Bluetooth Low Energy Separate Channel Fingerprinting

Shigemi Ishida, Yoko Takashima, Shigeaki Tagashira and Akira Fukuda

Abstract Bluetooth Low Energy (BLE) based localization is a next candidate for indoor localization. In this paper, we propose a new BLE-based fingerprinting localization scheme that improves localization accuracy. BLE is a narrow bandwidth communication that is highly affected by frequency selective fading. Frequency selective fading is mainly caused by multipaths between a transmitter and receiver, which are dependent on channels. We utilize channel specific features by separately measure received signal strength (RSS) on different channels to improve localization accuracy. BLE standards provide no API to retrieve channel information of incoming packets. We therefore developed a separate channel advertising scheme to separately measure RSS on different channels. To demonstrate the feasibility of the separate channel fingerprinting, we conducted preliminary experiments as well as initial evaluations. Experimental evaluations demonstrated that the separate channel fingerprinting improves localization accuracy by approximately 12%.

1 Introduction

Localization systems play an important role in location-based services. Location-based services in outdoor environment are now prevalent as smartphones equipped with a global positioning system (GPS) module are widely used.

A conference paper [11] containing preliminary results of this paper appeared in IIAI AAI ESKM 2016.

S. Ishida (✉) · Y. Takashima · A. Fukuda
ISEE, Kyushu University, Fukuoka 819-0395, Japan
e-mail: ishida@f.ait.kyushu-u.ac.jp

Y. Takashima
e-mail: takashima@f.ait.kyushu-u.ac.jp

A. Fukuda
e-mail: fukuda@f.ait.kyushu-u.ac.jp

S. Tagashira
Faculty of Informatics, Kansai University, Osaka 569-1095, Japan
e-mail: shige@res.kutc.kansai-u.ac.jp

© Springer International Publishing AG 2018
T. Matsuo et al. (eds.), *New Trends in E-service and Smart Computing*,
Studies in Computational Intelligence 742, https://doi.org/10.1007/978-3-319-70636-8_2

To extend location-based services to indoor environments, indoor localization is more required. Much effort has been paid to develop indoor localization schemes using technologies such as ultrasound, infrared light, and wireless signals. In particular, the localization scheme using Bluetooth Low Energy (BLE) is a next candidate for indoor localization. BLE is a low-power wireless technology suitable for battery-powered mobile devices. A large number of smartphones are equipped with Bluetooth modules that can transmit and receive BLE signals.

BLE-based localization systems, however, exhibit low localization accuracy because of both low transmission power and narrow bandwidth. Apple iBeacon is one of the famous localization systems using BLE and suffers from low accuracy. The iBeacon does not estimate an actual location but proximity to BLE beacons in three levels: within 10 cm as *immediate*, within one meter as *near*, and more than one meter as *far*. There is also an *unknown* status for failures.

BLE localization has been studied for range-based localization schemes [5, 10, 27] and a fingerprinting scheme [6]. In these works, BLE beacons are installed in an environment to broadcast advertising packets. A user device measures received signal strength (RSS) of the advertising packets to estimate own location. However, localization error becomes high up to approximately five meters because BLE uses three channels for advertising. Channel responses in the three advertising channels are often different because of frequency selective fading, resulting in the RSS instability without advertising channel information.

We present a new BLE-based localization scheme that improves localization accuracy. Our key idea is to employ channel specific features in a fingerprinting localization scheme. BLE uses narrow band channels and is highly affected by multipaths, which is dependent on channels and locations of both transmitter and receiver. We construct a fingerprint database that separately stores signal strength in three advertising channels to employ channel specific features.

BLE defines no API to retrieve channel information of received advertising packets. We therefore developed a separate channel BLE advertising scheme. BLE beacons periodically switch their transmission channels and embed transmission channel number in advertising packets. Note that some operating systems such as iOS above 7 optionally provide advertising channel information. We avoid to use such non-standard APIs to realize localization on all standards-compliant devices.

Specifically, our key contributions are threefold:

- We present a new BLE fingerprinting scheme that employs channel specific feature to improve localization accuracy. To the best of our knowledge, this is a first attempt to employ channel specific features for accuracy improvement in BLE fingerprinting localization.
- We present a separate channel advertising scheme in which BLE beacons periodically switch their transmission channel to transmit advertising packets including transmission channel information.
- We conduct experiments to confirm the BLE channel specific features and to demonstrate the feasibility of the separate channel fingerprinting scheme.

The remainder of this paper is organized as follows. Section 2 reviews related works on localization to clearly show our key contributions. We then present the design of BLE separate channel fingerprinting in Sect. 3, beginning with a key idea of our method. In Sect. 4, we conduct a preliminary experiment and an initial evaluation to demonstrate accuracy improvement by the separate channel fingerprinting. Finally, Sect. 5 concludes the paper.

2 Related Works

To the best of our knowledge, BLE fingerprinting separately employing fingerprints in three advertising channels is novel in the field of BLE localization. Much literature has been studied indoor localization. This section briefly looks through literature studying on indoor localization methods using radio signals in two categories: range-based and range-free localization methods. Most studies are primarily targeting WiFi localization systems but are applicable to other wireless technologies including BLE.

2.1 Range-Based Localization

Range-based localization methods estimate a location based on distances from anchor nodes whose locations are manually measured. A device to be localized communicates with anchor nodes and measures received signal strength (RSS) of packets from the anchor nodes to estimate the distances from the anchor nodes. The device then estimates its location by a localization method such as multilateration.

Studies on range-based localization methods have primarily investigated deployment cost reduction. Iterative multilateration reduces anchor deployment costs by utilizing localized nodes as new anchor nodes [16]. Iterative multilateration minimizes the number of initial anchor nodes because the number of anchor nodes increases as nodes are localized. Optimization of anchor locations also helps to reduce the number of initial anchor nodes [9]. These methods are helpful to reduce the number of BLE beacons in our localization system. Range-based localization using BLE, however, suffers from low accuracy compared to fingerprinting as they use unstable RSS of BLE signals. BLE devices handle incoming packets without channel information, resulting in instability of RSS on different channels caused by frequency selective fading.

Time of arrival (TOA), time difference of arrival (TDOA), and angle of arrival (AOA) methods can also be categorized into range-based localization methods as they rely on physical dimensions from anchor nodes. Standards-compliant BLE devices provide no API to retrieve these physical dimensions. These methods require a hardware extension.

2.2 Range-Free Localization

Pioneering range-free localization methods appear as localization systems for sensor and ad-hoc networks due to their simple calculations. Simple range-free localization methods such as Centroid [3], DV-hop [18], Amorphous [17], and APIT [7, 8] are based on network connectivity. These methods sacrifice localization accuracy for simple calculations and are impractical other than sensor networks.

Fingerprinting is another form of range-free localization and is widely used in indoor localization systems due to its high localization accuracy [1]. Fingerprinting consists of *learning* and *estimating* phases. A learning phase constructs a fingerprint database that stores fingerprints collected at everywhere in a target localization area. Fingerprint is a received signal strength (RSS) vector of beacon signals from anchor nodes installed in the environment, which is specific to the collected location. In an estimating phase, device location is estimated by comparing the RSS measured at the location with the fingerprints. The high accuracy of the fingerprinting is supported by a site-survey that collects enormous amounts of RSS at many locations in a target area.

Many studies on fingerprinting localization have reported accuracy improvements [12, 13, 15, 21, 24, 26]. Our separate channel fingerprinting can be applied in parallel to these methods to more improve localization accuracy.

Fingerprinting using the Bluetooth Classic utilizes inquiry processes to measure RSS [4, 25]. Bluetooth inquiry takes 5.12 seconds to discover 99% of devices [20]. The slow discovery makes difficult to realize Bluetooth localization systems in practical mobile scenarios.

Recent Bluetooth 4.0, i.e., BLE, addresses the slow discovery problem by employing a small number of channels for device discovery [2]. BLE uses three advertising channels to broadcast existence of BLE devices, which enables us to discover more quickly compared to the Bluetooth Classic.

Maximizing the advantage of the short discovery time, BLE fingerprinting was proposed in [6]. The study experimentally demonstrated RSS difference in three advertising channels. The difference is mainly caused by frequency selective fading as well as different channel gains of transmitter/receiver antennas and radio hardware. The study therefore finds channels highly affected by frequency selective fading and excludes the channels in fingerprint construction, minimizing the effect of channel difference. We are developing a fingerprinting scheme extending this study to utilize channel specific information to improve accuracy.

Because BLE localization suffers from low accuracy compared to WiFi localization, some studies have reported on accuracy improvement by such as collaborating BLE devices [22], combining room geometry information [14], employing weighted factor in fingerprint comparison [19], and optimizing BLE beacon locations [23]. These approaches are again parallel to our separate channel fingerprinting and contribute more improvement of accuracy.

3 Separate Channel Fingerprinting

3.1 Key Idea

Our key idea is to employ channel specific received signal strength (RSS) features in fingerprinting. Figure 1 shows distributions of RSS of advertising packets from an identical BLE beacon collected on three advertising channels at two locations 1 and 2. At each location, we collected advertising packets for 10 min. Red dotted lines in the figure indicate mean RSS. As shown in Fig. 1, RSS has different distributions on three advertising channels at a specific location with slightly different mean RSS. The distributions also depend on the measured location. We therefore utilize features such as mean and standard deviation of RSS as fingerprints to improve localization accuracy. We more precisely analyze RSS distributions in Sect. 4.1.

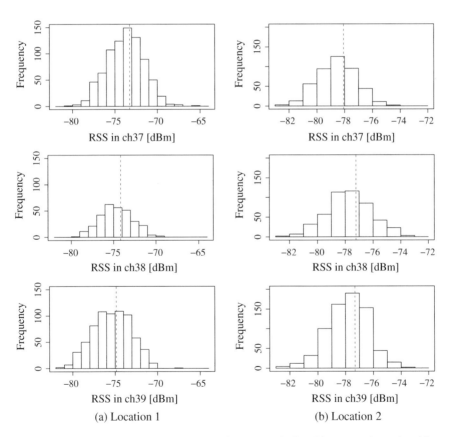

Fig. 1 Histograms of RSS of advertising packets from an identical BLE beacon on three advertising channels at two locations 1 and 2. Red dotted lines indicate mean RSS. RSS distributions are quite different with slightly different mean RSS. The RSS distribution is also dependent on the measured location

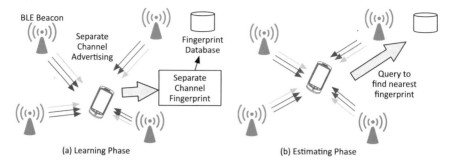

Fig. 2 System overview. **a** In a learning phase, a user device measures RSS of advertising packets in three channels and construct a separate channel fingerprint. The separate channel fingerprint is stored into a fingerprint database. **b** In an estimating phase, a user device measures RSS of advertising packets in three channels and looks up fingerprints from the database that is nearest to the measured RSS to estimate own location

3.2 System Overview

A BLE separate channel fingerprinting system consists of two components: separate channel advertising and fingerprinting using separate channel information. Figure 2 shows an overview of the BLE separate channel fingerprinting system. BLE beacons are installed in an environment beforehand. Each BLE beacon transmits advertising packets including transmission channel information in three advertising channels.

In a learning phase, we perform site-survey; the advertising packets are received by a BLE device at many locations in a target area to build a fingerprint database. The advertising packets from an identical beacon on different advertising channels are handled as if there are three BLE beacons operating in three different channels. The BLE device therefore groups advertising packets by sender BLE beacons and advertising channels to calculate separate channel fingerprints. The separate channel fingerprints are stored in a fingerprint database.

In an estimating phase, a user BLE device receives advertising packets and measures the RSS of the packets to calculate a separate channel fingerprint. The user device then sends query to a fingerprint database to look up the nearest fingerprint to the separate channel fingerprint calculated at the location to estimate device location.

Following subsections describe details of the each component.

3.3 Separate Channel Advertising

Figure 3 illustrates an overview of separate channel advertising. In a separate channel advertising scheme, BLE beacons apply a mask to limit their advertising channels. BLE beacons embed their advertising channel information into advertising packets because BLE provides no API to retrieve channel information of received pack-

Fig. 3 Overview of separate channel advertising. A BLE beacons periodically switches a mask to limit its advertising channels. BLE defines no API to retrieve channel information of received packets. We therefore embed advertising channel information in advertising packets

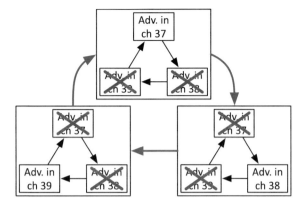

Fig. 3 Overview of separate channel advertising. A BLE beacons periodically switches a mask to limit its advertising channels. BLE defines no API to retrieve channel information of received packets. We therefore embed advertising channel information in advertising packets

ets. User devices that receive an advertising packet decode the embedded data to retrieve transmission channel information. The channel mask is periodically updated to change an advertising channel, as shown in Fig. 3.

We implemented BLE beacons that send advertising packets compatible with Apple iBeacon. BLE beacons embed advertising channel information as 2-bit data into a `Minor` field in an iBeacon advertising packet.

Although some BLE modules provide a vendor specific host control interface (HCI) to limit transmission channels, advertising channel mask is not defined in a BLE standards specification. We need to use specific BLE beacon hardware, which is a main limitation of our scheme.

3.4 Fingerprinting Utilizing Separate Channel Information

Fingerprinting utilizing separate channel information consists of learning and estimating phases, which is the same as conventional fingerprinting consists of learning and estimating phases.

In a learning phase, we collect separate channel fingerprints at many locations in a target area. We divide localization area into small sub-areas and measure RSS of BLE beacons in the each sub-area. At each sub-area, we collect advertising packets for a specific duration. Let i denote a sub-area and n denote the number of BLE beacons. Location fingerprint R_i in a sub-area i is a $3n$-th vector:

$$R_i = \{\overline{r_{i1_{37}}}, \overline{r_{i1_{38}}}, \overline{r_{i1_{39}}}, \overline{r_{i2_{37}}}, \ldots, \overline{r_{in_{39}}}\}, \tag{1}$$

where $\overline{r_{ij_c}}$ ($j = 1, 2, \ldots, n$ and $c = 37, 38, 39$) is a mean RSS of advertising packets from a BLE beacon j in a sub-area i on an advertising channel c. We used $-\infty$ as an RSS of undetected BLE beacons. We also calculate a standard deviation vector σ_i for all the BLE beacons j ($j = 1, 2, \ldots, n$) in each sub-area i on an advertising channel c ($c = 37, 38, 39$) in the same manner:

$$\sigma_i = \{\sigma_{i1_{37}}, \sigma_{i1_{38}}, \sigma_{i1_{39}}, \sigma_{i2_{37}}, \ldots, \sigma_{in_c}\}, \tag{2}$$

where σ_{ij_c} is a standard deviation of RSS of a BLE beacon j in a sub-area i on an advertising channel c.

In an estimating phase, we estimate location of a user device based on distance between fingerprints. A user device measures RSS of advertising packets and calculate a separate channel fingerprint $x = \{\overline{x_{1_{37}}}, \overline{x_{1_{38}}}, \ldots, \overline{x_{n_{39}}}\}$ in the same manner as Eq. (1). The user device next calculates distance between the calculated fingerprint x and the fingerprints R_i collected in a learning phase. We used ℓ^1 norm for distance calculation. The device location i is finally estimated as

$$i = \arg\min_i distance(R_i, x), \tag{3}$$

$$\text{where } distance(R_i, x) = \frac{1}{3n} \sum_{j=1}^{n} \sum_{c \in \{37,38,39\}} |\overline{r_{ij_c}} - \overline{x_{j_c}}|.$$

Undetected BLE beacons described in $-\infty$ are ignored in distance calculation.

To avoid huge localization errors, we apply a simple filter using the standard deviation vector σ_i. Device location is estimated as *unknown* when there is one or more $\overline{x_{j_c}}$ that satisfies $|\overline{r_{ij_c}} - \overline{x_{j_c}}| > \sigma_{ij_c}$.

4 Initial Evaluation

To demonstrate the feasibility of the BLE separate channel fingerprinting described in Sect. 3, we evaluated accuracy of estimated locations. We first conducted a preliminary experiment to confirm the difference of RSS distributions on three advertising channels. We then installed a small number of BLE beacons and estimated a small number of locations as an initial evaluation.

4.1 Preliminary Experiment

A preliminary experiment was conducted to confirm the difference of RSS distributions on three advertising channels. We installed a Bluegiga BLED112 USB dongle performing separate channel advertising in our laboratory. We then collected BLE advertising packets using a MacBook Air laptop for 10 min at four locations and compared RSS distributions on three advertising channels.

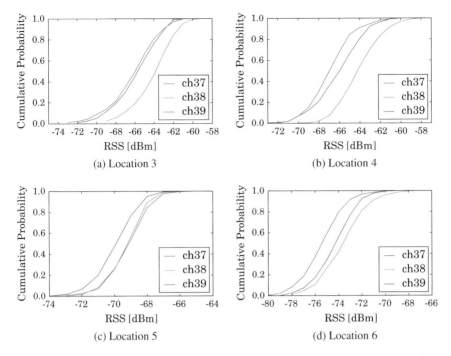

Fig. 4 Empirical cumulative distribution functions (ECDFs) of RSS of advertising packets from an identical BLE beacon on three advertising channels at four locations 3, 4, 5, and 6

Figure 4 shows empirical cumulative distribution functions (ECDFs) of RSS of advertising packets from an identical BLE beacon on three advertising channels at four locations. Figure 4 shows the following:

- At each location, RSS of advertising packets exhibited different distributions on three advertising channels. BLE is a narrow bandwidth communication technology and is highly affected by frequency selective fading. The frequency selective fading causes different channel gains on the three channels, which resulted in different RSS distributions.
- The RSS distribution difference was dependent on locations. The frequency selective fading is mainly caused by multipaths, which depend on locations of both transmitter and receiver.

To confirm the RSS difference, we performed two-sample t-tests on the RSS of advertising packets. Table 1 shows the p values obtained by the Welch's two-sample t-tests on the RSS. Each cell in the table represents a p value for two channels and locations indicated by the row and column. Gray colored cells are non-zero cells. In Table 1, almost all the cells are zero. At a significance level of $p < 0.05$, we can confirm that almost all the channel combinations exhibited significant difference except three channel combinations: channel 37 at location 3 and channel 39 at location 4, channel 39 at locations 3 and 4, and channels 37 and 38 at location 5.

Table 1 The results of Welch's two-sample t-tests on the RSS of advertising packets received at different locations and advertising channels. Each cell represents a p value for two channels indicated by the row and column. Gray colored cells are non-zero cells. Almost all the cells are zero, which indicates the mean RSS values on these channels and locations were significantly different

		Location 3			Location 4			Location 5			Location 6		
		ch37	ch38	ch39	ch37	ch38	ch39	ch37	ch38	ch39	ch37	ch38	ch39
Location 3	ch37	–	0.000	0.026	0.000	0.000	0.609	0.000	0.000	0.000	0.000	0.000	0.000
	ch38	0.000	–	0.000	0.000	0.000	0.000	0.000	0.000	0.000	0.000	0.000	0.000
	ch39	0.026	0.000	–	0.000	0.000	0.082	0.000	0.000	0.000	0.000	0.000	0.000
Location 4	ch37	0.000	0.000	0.000	–	0.000	0.000	0.000	0.000	0.000	0.000	0.000	0.000
	ch38	0.000	0.000	0.000	0.000	–	0.000	0.000	0.000	0.000	0.000	0.000	0.000
	ch39	0.609	0.000	0.082	0.000	0.000	–	0.000	0.000	0.000	0.000	0.000	0.000
Location 5	ch37	0.000	0.000	0.000	0.000	0.000	0.000	–	0.344	0.000	0.000	0.000	0.000
	ch38	0.000	0.000	0.000	0.000	0.000	0.000	0.344	–	0.000	0.000	0.000	0.000
	ch39	0.000	0.000	0.000	0.000	0.000	0.000	0.000	0.000	–	0.000	0.000	0.000
Location 6	ch37	0.000	0.000	0.000	0.000	0.000	0.000	0.000	0.000	0.000	–	0.000	0.000
	ch38	0.000	0.000	0.000	0.000	0.000	0.000	0.000	0.000	0.000	0.000	–	0.000
	ch39	0.000	0.000	0.000	0.000	0.000	0.000	0.000	0.000	0.000	0.000	0.000	–

Fig. 5 Experiment setup. Greek labels α–δ indicate BLE beacon locations and alphabetical labels A–G indicate receiver locations to be estimated. We installed four Bluegiga BLED112 USB dongles and measured RSS of advertising packets from the each BLE dongle for approximately 10 min using BLE receiver on MacBook Pro

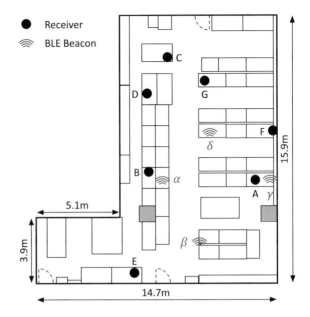

4.2 Experiment Setup

Figure 5 depicts an experiment setup. We used Bluegiga BLED112 USB dongles as BLE beacons. Four BLE beacons were installed in our laboratory at locations having Greek labels in Fig. 5. Each BLE beacon transmitted advertisement packets every 100 milliseconds and switched its advertising channel every 300 milliseconds, as described in Sect. 3.3. We note that there were 20 WiFi APs operating in a 2.4-GHz band in and around our laboratory.

We collected advertising packets from each BLE beacon at locations labeled A and B to measure received signal strength (RSS) for approximately 60 min using two MacBook Pros. At the same time, we also measured RSS at locations labeled C to G for approximately 10 min at each location using another MacBook Pro.

Using the collected RSS data, we estimated locations within A, B, and unknown. Estimated locations were evaluated using a ten-fold cross-validation. As shown in Fig. 6, we divided the RSS data collected at locations A and B into ten chunks. Let N be the number of $1/10$ RSS data. The $9N$ data was used to construct a fingerprint database and the N data was used as an input of a location estimator. The N data at locations C to G were also used as an input of the location estimator, which should be estimated as unknown.

Comparing estimated locations with actual locations, we evaluated the number of true positives (TPs), false negatives (FNs), false positives (FPs), and true negatives (TNs). TPs, FNs, FPs, and TNs are defined as the case that locations were correctly estimated, locations were mistakenly estimated as unknown, locations were mistakenly estimated as A or B, and locations were correctly estimated as unknown,

Fig. 6 Data usage in
ten-fold cross-validation.
Data is divided into ten
chunks. The 9/10 of the data
was used for learning and the
remaining 1/10 was used for
estimation

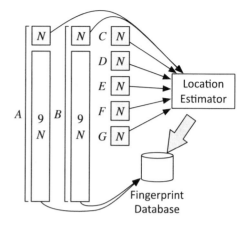

respectively. Using the number of TPs, FNs, FPs, and TNs, we also evaluated an
accuracy, precision, recall, and F-measure defined as:

$$\text{Accuracy} = \frac{\text{TP} + \text{TN}}{\text{TP} + \text{FP} + \text{FN} + \text{TN}}, \tag{4}$$

$$\text{Precision} = \frac{\text{TP}}{\text{TP} + \text{FP}}, \tag{5}$$

$$\text{Recall} = \frac{\text{TP}}{\text{TP} + \text{FN}}, \tag{6}$$

$$\text{F}_{\text{measure}} = \frac{2 \cdot \text{Precision} \cdot \text{Recall}}{\text{Precision} + \text{Recall}}. \tag{7}$$

4.3 Location Estimation Accuracy

Table 2 shows experiment results. The total number of estimations was 3988 times.
Table 2a shows that the separate channel fingerprinting increased the number of TPs
while reducing the number of FPs. We can also confirm that the number of FNs was
also reduced.

Table 2b shows that an accuracy, precision, recall, and F-measure were increased
by separate channel fingerprinting. Accuracy was improved by $(0.55 - 0.49)/0.49 \times 100 \simeq 12\%$. We can conclude that the separate channel fingerprinting has the capa-
bility to improve localization accuracy.

Accuracies of both conventional and proposed methods, however, were approximately 0.5, which was too low for practical applications. The main reason of the low
accuracies was the large number of FPs. In our experiment, locations where a fingerprint was distant from any fingerprints in a database were estimated as `unknown`.
Fingerprinting schemes provide no considerations on location estimation outside of
target areas, which resulted in the large number of FPs.

Table 2 Location estimation results. Proposed separate channel fingerprinting increased the number of TPs, while reducing the number of both FPs and FNs compared to conventional fingerprinting. Accuracy was improved by approximately 12%

(a) Number of TPs, FNs, FPs, and TNs

	TPs	FNs	FPs	TNs
Conventional	1117	59	1977	836
Proposed	1319	30	1775	864

(b) Accuracy, precision, recall, and F-measure

	Accuracy	Precision	Recall	F-measure
Conventional	0.49	0.36	0.95	0.52
Proposed	0.55	0.43	0.98	0.59

5   Conclusion

In this paper, we proposed a BLE separate channel fingerprinting that employs channel specific features in fingerprinting to improve localization accuracy. BLE standards provide no API to recognize advertising channels. We therefore developed a separate channel advertising scheme, which enables standards-compliant BLE devices to recognize advertising channels. By conducting a preliminary experiment, we confirmed the signal strength difference on different advertising channels. Initial evaluations demonstrated that the separate channel fingerprinting improves location estimation accuracy by approximately 12%. We are now working on development of BLE localization system utilizing the separate channel fingerprinting.

Acknowledgements This work was supported in part by JSPS KAKENHI Grant Number 15H05708 and the Cooperative Research Project of the Research Institute of Electrical Communication, Tohoku University.

References

1. Bahl, P., Padmanabhan, V.N.: RADAR: an in-building RF-based user location and tracking system. In: Proceedongs of the IEEE International Conference on Computer Communications (INFOCOM), pp. 775–784 (2000)
2. Bluetooth Special Interest Group: Core specification v4.0 (2010). http://www.bluetooth.com/
3. Bulusu, N., Heidemann, J., Estrin, D.: GPS-less low-cost outdoor localization for very small devices. IEEE Pers. Commun. Mag. 7(5), 28–34 (2000)
4. Chen, L., Pei, L., Kuusniemi, H., Chen, Y., Kröger, T., Chen, R.: Bayesian fusion for indoor positioning using Bluetooth fingerprints. Int. J. Wirel. Pers. Commun. 70(4), 1735–1745 (2013)
5. Contreras, D., Castro, M., de la Torre, S.: Performance evaluation of Bluetooth Low Energy in indoor positioning systems. Trans. Emerg. Telecommun. Technol. 25(8), 1–10 (2014)
6. Faragher, R., Harle, R.: Location fingerpriting with Bluetooth Low Energy beacons. IEEE J. Sel. Areas Commun. 33(11), 2418–2428 (2015)
7. He, T., Huang, C., Blum, B.M., Stankovic, J.A., Abdelzaher, T.: Range-free localization schemes for large scale sensor networks. In: Proceedings of the 9th Annual International Conference on Mobile Computing and Networking, pp. 81–95 (2003)
8. He, T., Huang, C., Blum, B.M., Stankovic, J.A., Abdelzaher, T.F.: Range-free localization and its impact on large scale sensor networks. ACM Trans. Embed. Comput. Syst. (TECS) 4(4), 877–906 (2005)
9. Huang, L., Wang, F., Ma, C., Duan, W.: The analysis of anchor placement for self-localization algorithm in wireless sensor networks. In: Advances Wireless Sensor Networks, Communications in Computer and Info. Science, vol. 334, pp. 117–126 (2013)
10. Ionescu, G., de la Osa, C.M., Deriaz, M.: Improving distance estimation in object localisation with Bluetooth Low Energy. In: Proceedings of the International Conference on Sensor Technologies and Applications (SENSORCOMM), pp. 1–5 (2014)
11. Ishida, S., Takashima, Y., Tagashira, S., Fukuda, A.: Proposal of separate channel fingerprinting using Bluetooth Low Energy. In: Proceedings of the IIAI International Congress Advanced Applied Informatics (AAI), ESKM, pp. 230–233 (2016)
12. Kaemarungsi, K., Krishnamurthy, P.: Analysis of WLAN's received signal strength indication for indoor location fingerprinting. Pervasive Mob. Comput. 8(2), 292–316 (2012)
13. Kushki, A., Plataniotis, K.N., Venetsanopoulos, A.N.: Intelligent dynamic radio tracking in indoor wireless local area networks. IEEE Trans. Mobile Comput. 9(1), 405–419 (2010)

14. Kyritsis, A.I., Kostopoulos, P., Deriaz, M., Konstantas, D.: A BLE-based probabilistic room-level localization method. In: Proceedings of the International Conference on Localization and GNSS (ICL-GNSS), pp. 1–6 (2016)

15. LaMarca, A., Chawathe, Y., Consolvo, S., Hightower, J., Smith, I., Scott, J., Sohn, T., Howard, J., Hughes, J., Potter, F., Tabert, J., Powledge, P., Borriello, G., Schili, B.: Place lab: device positioning using radio beacons in the wild. In: LNCS, vol. 3468, pp. 116–133 (2005). Proceedings of the ACM International Conference on Pervasive Computing (PERVASIVE)

16. Minami, M., Fukuju, Y., Hirasawa, K., Yokoyama, S., Mizumachi, M., Morikawa, H., Aoyama, T.: DOLPHIN: A practical approach for implementing a fully distributed indoor ultrasonic positioning system. In: LNCS, vol. 3205, pp. 437–365 (2004). Proceedings of the ACM Conference Ubiquitous Computing (Ubicomp)

17. Nagpal, R., Shrobe, H., Bachrach, J.: Organizing a global coordinate system from local information on an ad hoc sensor network. In: LNCS, vol. 2634, pp. 333–348 (2003). Proceedings of the IPSN

18. Niculescu, D., Nath, B.: Ad hoc positioning system (APS). In: Proceedings of the IEEE GLOBECOM, pp. 2926–2931 (2001)

19. Peng, Y., Fan, W., Dong, X., Zhang, X.: An iterative weighted KNN (IW-KNN) based indoor localization method in Bluetooth Low Energy (BLE) environment. In: Proceedings of the IEEE International Conference on Cloud and Big Data Computing (CBDCom), pp. 794–800 (2016)

20. Peterson, B.S., Baldwin, R.O., Kharoufeh, J.P.: Bluetooth inquiry time characterization and selection. IEEE Trans. Mob. Comput. 5(9), 1173–1187 (2006)

21. Prasithsangaree, P., Krishnamurthy, P., Chrysanthis, P.K.: On indoor position location with wireless LANs. In: Proceedings of the IEEE International Symposium on Personal, Indoor and Mobile Radio Communications (PIMRC), pp. 720–724 (2002)

22. Qui, J.W., Lin, C.P., Tseng, Y.C.: BLE-based collaborative indoor localization with adaptive multi-lateration and mobile encountering. In: Proceedings of the IEEE International Wireless Communications and Networking Conference, pp. 1–7 (2016)

23. Schmalenstroeer, J., Haeb-Umbach, R.: Investigations into Bluetooth Low Energy localization precision limits. In: Proceedings of the European Signal Processing Conference (EUSIPCO), pp. 652–656 (2016)

24. Sen, S., Radunović, B., Choudhury, R.R., Minka, T.: You are facing the Mona Lisa: Spot localization using PHY layer information. In: Proceedings of the ACM MobiSys, pp. 183–196 (2012)

25. Subhan, F., Hasbullah, H., Rozyyev, A., Bakhsh, S.T.: Indoor positioning in Bluetooth networks using fingerprinting and lateration approach. In: Proceedings of the IEEE International Conference on Information Science Applications (ICISA), pp. 1–9 (2011)

26. Tsui, A.W., Chuang, Y.H., Chu, H.H.: Unsupervised learning for solving RSS hardware variance problem in WiFi localization. Mob. Netw. Appl. 12(5), 677–691 (2009)

27. Zhu, J., Chen, Z., Luo, H., Li, Z.: RSSI based Bluetooth Low Energy indoor positioning. In: Proceedings of the International Conference on Indoor Positioning and Indoor Navigation (IPIN), pp. 526–533 (2014)

Toward Sustainable Smart Mobility Information Infrastructure Platform: Project Overview

Akira Fukuda, Kenji Hisazumi, Tsunenori Mine, Shigemi Ishida,
Takahiro Ando, Shota Ishibashi, Shigeaki Tagashira, Kunihiko Kaneko,
Yutaka Arakawa, Weiqiang Kong and Guoqiang Li

Abstract Smart mobility systems, which include Intelligent Transportation Systems (ITS) and smart energy ones, become more important. There is, however, lack of its platform studies. This paper proposes a sustainable information infrastructure project for smart mobility systems. The project pursues issues that establish an information infrastructure architecture and a seamless development method chain for it. The project has mainly two features: (1) applying life-cycle-oriented methods, which are a cycle from system development phase to operation phase. In addition, these methods are applied to real world applications, and (2) dealing with an uncertainty that occurs in system development upper phase. This paper describes an overview of the project.

A. Fukuda (✉) · K. Hisazumi · T. Mine · S. Ishida · T. Ando · S. Ishibashi
Faculty of Information Science and Electrical Engineering,
Kyushu University, Fukuoka, Japan
e-mail: fukuda@ait.kyushu-u.ac.jp

K. Hisazumi
e-mail: nel@slrc.kyushu-u.ac.jp

T. Mine
e-mail: mine@ait.kyushu-u.ac.jp

S. Ishida
e-mail: ishida@f.ait.kyushu-u.ac.jp

T. Ando
e-mail: ando.takahiro@f.ait.kyushu-u.ac.jp

S. Ishibashi
e-mail: ishibashi@f.ait.kyushu-u.ac.jp

S. Tagashira
Faculty of Informatics, Kansai University, Suita, Japan
e-mail: shige@res.kutc.kansai-u.ac.jp

K. Kaneko
Faculty of Engineering, Fukuyama University, Fukuyama, Japan
e-mail: kaneko@fuip.fukuyama-u.ac.jp

Y. Arakawa
Graduate School of Information Science, Nara Institute of Science
and Technology, Ikoma, Japan
e-mail: ara@is.naist.jp

© Springer International Publishing AG 2018   35
T. Matsuo et al. (eds.), *New Trends in E-service and Smart Computing*,
Studies in Computational Intelligence 742, https://doi.org/10.1007/978-3-319-70636-8_3

1 Introduction

Recently, realizing smart society becomes one of hot topics. In particular, as one of smart societies, smart mobility, which includes Intelligent Transportation System (ITS) and smart energy systems, becomes more important. This is because the world is shifted to human familiar rather than technology oriented to resolve human mobility problems.

There are lots of researches for smart mobility. Most of these researches, however, have focused each separated techniques such as sensing acquisition, visualization of automobile probe data, etc. In addition, there are many platforms for smart mobility. However, these platforms are closed and separated from each other smart mobility fields. There is a lack of systematic/foundational research on system design, development, and construction of an information infrastructure for smart mobility over entire fields. For the infrastructure research, it is important that life-cycle-oriented methods must be constructed, since smart mobility systems rapidly change. The life-cycle-oriented methods mean ones that are applied from system design and development to system operations and feed-backed from the operations to the development after performing operations. This cycle allows us to meet the requirements through performing operations into system design and development.

This paper proposes a sustainable smart mobility information infrastructure project [1], which are applied to many smart mobility fields. The project pursues to establish an information infrastructure architecture and a seamless-development-methods-chain for it. The project has mainly two features: (1) applying life-cycle-oriented methods, which are a cycle from system development to operations, to real world applications, and (2) dealing with uncertainty when some items are not clearly decided in an upper phase of a system development. This paper describes an overview of the project.

The paper is structured as follows. A smart mobility information infrastructure is described in Sect. 2. Our project overview including its features and challenging issues is described in Sect. 3. Handling uncertainty, which is one of our project's features, is given in Sect. 4. Our current preparations for the project are described in Sect. 5. Some issues of the project are discussed in Sect. 6. Related works are described in Sect. 7. Finally, conclusion and future work are discussed in Sect. 8.

W. Kong
School of Software, Dalian University of Technology, Dalian, China
e-mail: wqkong@dlut.edu.cn

G. Li
School of Software, Shanghai Jiao Tong University, Shanghai, China
e-mail: li.g@sjtu.edu.cn

Fig. 1 Smart mobility information infrastructure platform

2 Smart Mobility Information Infrastructure Platform

(1) Platform for smart mobility

The smart mobility information infrastructure platform is shown in Fig. 1. It consists of three components: information collection into a platform, the platform, and information provision services.

1. The information collection component:

The information collection component gathers data from sensing physical world, which includes, e.g., new sensors, shop or tourist resort information, government information, original ITS information, various services information by vendors, etc., that is, IoT (Internet of Things). These data collected are into the platform as big data.

2. The platform component:

The platform component consists of the following.

Many assets from system design and development upper phase to lower one, which include system documents, design specification, source code programs, etc. The platform allows business companies to provide new services to users.

3. The information provision service component:

By using the platform, business companies can provide various innovative services to users. These services include normal or vulnerable road ones and smart mobility oriented novel ones.

(2) Scope [2]

In the project, smart mobility as a service is defined to optimize or adjust interests among stakeholders related to the transportation including human beings as well as things. Smart mobility should be human centric.

There are four kinds of stakeholders [2]:

Users: Ones who use public/private transportation by using information provided from service providers.

Public/private transportation providers: Ones who provide transportation such as planes, ships, railways, buses, taxis, ridesharing cars, rent-a-cars, private cars, and walking, etc.

Service providers: Ones who provide information to the users and the transportation providers to optimize or adjust transportation means by considering the stakeholders.

Government/Local Governments: Ones who approve transportations, to control them in terms of total optimization according to laws and taxes.

The project must consider these stakeholders.

The project assumes the architecture for smart mobility as follows, as shown in Fig. 2:

The Platform: is a kind of middleware that supports applications to provide functions to manage information for smart mobility. It supports to obtain information from sensors, and to send it to actuators. The platform allows system operations in applications to be easily handled.

Applications: provide information to the users and public/private transportation providers to optimize or adjust transportation means by considering the stakeholders. Applications get data from sensors and send information to actuators through the platform. The applications run in a cloud environment, smartphones, embedded systems, etc.

Sensors: sense the real world and get its state. The sensors can be embedded in the applications or be independent from each other.

Actuators: actuate some devices or machines to change the state of real world. The actuators can be embedded in the applications.

Fig. 2 The architecture using the smart mobility platform

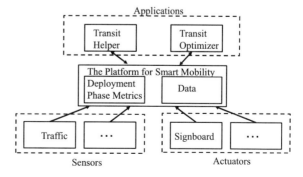

Fig. 3 Life-cycle-oriented development

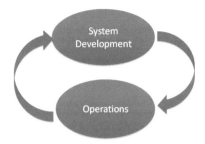

3 Project Overview

(1) Features of the project

Our project has the following features.

1. Real Life-Cycle Oriented Project

The smart mobility systems including ITS and smart energy systems are rapidly changing. Therefore, it is important to rapidly construct a system and to be feed-backed to the system design and development after operations. Many researches and projects that deal with life-cycle system development have been done (Fig. 3). Most of these researches, however, have not applied to real world fields. Our project is for real fields. The proposed methods in the project will be applied to real world fields. In particular, we have a plan to invite the biggest ITS company in Japan to join the project. This could guarantee our project's success.

2. Dealing with Uncertainty

The most important feature is that our project deals with uncertainty.

Recently, many systems become big and complicated. In addition, its user requirements become not clear in system development upper phase. Some parts of the system itself and operation ways are not decided in the upper phase, that is, uncertainty. These parts are decided in lower phases of system development or operation phase.

(2) Challenging issues of the project

The object of our project is to establish a sustainable information platform architecture for smart mobility systems. The project pursues the following challenging issues, as shown in Fig. 4.

Establishing life-cycle-oriented architecture

Life-cycle-oriented architecture that includes a cycle from system development and construction to operations seems to be not clear so far. Therefore, the project proposes its architecture.

Establishing life-cycle-oriented technologies

Issues that occur in operation phase after system construction must be feed-backed to the system development. Therefore, the project develops traceability technologies that are from the operations to system development.

Fig. 4 Challenging issues of
the project

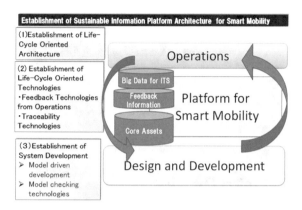

Establishing system development in upper phase

In system development process, its upper phases such as system modelling and checking phase are very important. For these phases, many researches have been done. The project also develops model-driven-based techniques and model checking ones, and applies them to real fields.

Establishing Seamless System Development Cycle

System development methods have been studied so far. Most of these methods, however, seem to be disjoined from each other, which is not seamless. Seamless cycle that joins each method is needed. Our project develops a seamless process.

4 Handling the Uncertainty

Uncertainty including its definition, categorization, etc. in the project is described in some references. The definition and the categorization is described in Ref [3]. Reference [4] described an analysis of uncertainty combined with DevOps and integrated into dynamic software product line process.

An outline of handling the uncertainty is as follows [5], as shown in Fig. 5:

(1) Step 1: Making an uncertainty list sheet

From system documents including requirements and specifications, we make an uncertainty list sheet. In the sheet, we make functions such as classifying the uncertainty, its impact factors [6], etc.

(2) Step 2: Making a case list

We extract items of the uncertainty from the sheet made in Step 1. In addition, we make a list that describes variability and its scope for each item.

(3) Step 3: Making a feature model

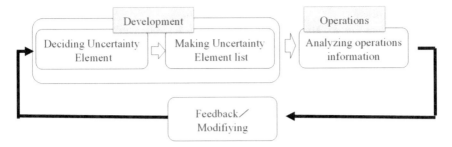

Fig. 5 Handling process of uncertainty between system development and operations

We employ a feature-oriented approach from the requirement. This approach allows us to make a feature model. We mark items of the uncertainty in the feature model. From this model, we can clarify the uncertainty in the target system.

(4) Step 4: Making a partial model including the uncertainty items.

We make a partial model that covers all uncertainty items marked by Step 4.

(5) Step 5: Changing the model when the uncertainty items are fixed.

When one of the uncertainty items is selected in the system development upper phase, we change the model to reflect it.

5 Preparation

5.1 Overview

Our project has performed many studies towards smart mobility as follows.

(1) Uncertainty

Uncertainty has been addressed in management and social sciences rather than in software engineering field so far. Our project has proposed a framework to manage uncertainty in software system development [2–6].

(2) Traceability tool for model-based development dealing with uncertainties.

The project has developed a traceability tool for model-based development dealing with uncertainties. This tool has features including meta-indexed [7].

(3) Model driven development techniques

The project has developed a model driven tool called Clooca [8]. The Clooca tool is web-based, which means that the tool is not installed for individual persons' computers. The tool is used in education field of Japanese universities.

(4) Model checking techniques

Many model checking techniques have been developed [9–13]. Our project has also developed the techniques and its tool called Garakabu2 [13]. Garakabu2 is provided as business tool by a Japanese company. In Sect. 5.2 in this paper, the model checking techniques that we developed are described.

(5) Applications

Our project has developed some applications including ITS field, on-demand-localization services, smart city, etc. [14–20].

5.2 Model Checking Technologies

As an example, we describe our preparation on model checking techniques in this subsection. Our work on model checking can be classified into algorithm development and tool implementation.

Regarding algorithm development, we have proposed a hybrid multicore algorithm, in which bounded model checking (BMC) [21] as well as multicore processing is guided by stateless explicit state-space exploration and related abstraction techniques. An overview of the algorithm is shown in Fig. 6. In this algorithm, stateless explicit state exploration techniques are first applied to explore the original state space and memorize legal execution paths. During exploration, abstraction techniques like Bounded Context Switch (BCS) can be applied to reduce the original state space; those memorized paths are next classified into path clusters based on systematically-generated heuristic predicates; each path cluster is then encoded into a separate propositional formula, which, together with an encoded formula for an LTL property, represents an independent BMC instance; the BMC instances are finally solved with multiple copies of an SMT solver that run on multicores. State exploration and BMC progress in an incremental manner and each iteration deepens the search bound with a pre-specified number.

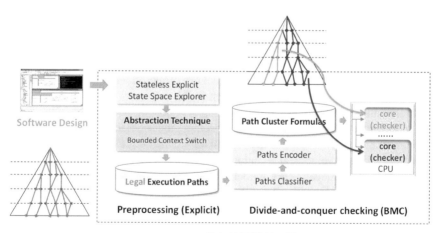

Fig. 6 An overview of our proposed hybrid multicore algorithm

Regarding tool implementation, we have implemented the above algorithm into a tool called Garakabu2. Garakabu2 accepts input models developed with State Transition Matrix modeling language in ZIPC environment [22], a popular language and case tool in Japan for embedded software development industry. We have conducted experiments on sets of benchmarks to evaluate performance of our algorithm and tool. The results show that Garakabu2 often outperforms the BMC algorithms implemented in well-known SAL (the same type of model checker as Garakabu2) and partially performs better than NuSMV. In addition to performance, we have also payed efforts to improve the usability of Garakabu2 for on-site software engineers, which include a push-button interface, graphical counterexample (co)simulation, and LTL specification assistance, etc.

6 Discussion

This paper described our project that performs sustainable smart mobility information infrastructure platform study. The project has mainly two features: (1) constructing a life-cycle-oriented information infrastructure platform from system design and development to operations, and (2) dealing with uncertainty that occurs in system development upper phase.

(1) Life-cycle-oriented development:

Recently, collaboration between system development and operations has been addressed. One of them is called DevOps (= Development + Operation). DevOps seems to be not clear in technologies. The project proposes its possible one of research directions in technology field.

(2) Uncertainty:

Uncertainty has been addressed in management field and social science one. Our project applies uncertainty into software development. Although uncertainty in the first stage of the project is limited, we make efforts to expand it.

(3) Seamless development cycle

Many software development technologies have been developed. These technologies, however, seem to be separated from each other. Our project pursues seamless techniques and these tools.

7 Related Work

(1) Uncertainty

In general describing, uncertainty seems to have been discussed much in the field of management science more than in the field of information engineering science. The term "uncertainty" has been used with confusion. Wynne classified the meanings of the term "uncertainty" into the following seven categories [23].

(1) Risk: known damage and probabilities;

(2) Uncertainty: known damage possibilities but no knowledge of probabilities;

(3) Ignorance: unknown unknowns ("second order uncertainty" Smithson);

(4) Indeterminacy: issue and conditions, hence knowledge-framing open; maybe salient behavioral processes also non-determinate;

(5) Complexity: open behavioral systems, and multiplex, often non-linear processes so that extrapolation from robust data-points always problematic;

(6) Disagreement: divergence over framing, observation methods or interpretation. Questions of competence of parties.

(7) Ambiguity: precise meanings (hence salient elements) not agreed, or unclear.

The uncertainty discussed in this paper corresponds to the risk and the uncertainty in the above Wynne's categories.

(2) DevOps (=Development + Operations)

Recently, DevOps, which means that collaboration between system development and its operations must be done, has been addressed. DevOps is defined as "a set of practices intended to reduce the time between committing a change to a system and the change being placed into normal production, while ensuring high quality" [24]. There are some studies. For example, Weiyi Shang proposed that gaps between software developers and operators must be resolved by using logs [25].

8 Conclusion

This paper proposed a sustainable smart mobility information infrastructure platform. In addition, the authors launched its project. The project will give an innovative research direction. The project has mainly two features, (1) applying life-cycle oriented system development methods to real world, and (2) dealing with uncertainty in system design and development upper phase. There are still, however, many issues that should be investigated to successfully accomplish the project. First, system evaluation performed by the project must provide useful information to information society. Second, the seamless tools developed must be provided to open-source ones for system developers and operators. Finally, the project must give a way to provide our project results for world.

Acknowledgements The project described in this paper is supported by JSPS KAKENHI Grant-in-Aid for Scientific Research(S), Grant Number 15H05708.

References

1. Fukuda, A., Nakanishi, T., Hisazumi, K., Tagashira, S., Arakawa, Y., Ishida, S., Mine, T., Kaneko, K., Furusho, H., Kong, W.: Towards sustainable information infrastructure platform for smart mobility-project overview. In: Proceeding the 5th IIAI International Congress on Advanced Applied Informatics (IIAI AAI 2016), Special Session ESKM, pp. 211–214 (2016)

2. Hisazumi, K., Nakanishi, T., Ishibashi, S., Hirakawa, G., Mine, T., Ando, T., Furusho, H., Fukuda, A.: Operation phase metrics for smart mobility platform. In: Proceeding IEEE International Conference on Agents (IEEE ICA 2016), Workshop of Sustainable Smart Mobility Platform (SSMP) (2016)
3. Nakanishi, T., Ma, L.-D., Hisazumi, K., Fukuda, A.: A Framework to Manage Uncertainty in System Development, IPSJ SIG Technical Report, Vol. 2014-SLDM-165, No. 6, 6 p (2014) (in Japanese)
4. Nakanishi, T., Furusho, H., Hisazumi, K., Fukuda, A.: Dynamic SPL and derivative development with uncertainty management for DevOps. In: Proceeding the 5th IIAI International Congress on Advanced Applied Informatics (IIAI AAI 2016), Special Session ESKM, pp. 244–249 (2016)
5. Ishibashi, S., Hisazumi, K., Nakanishi, T., Fukuda, A.: Establishing traceability between requirements, design and operation information in lifecycle-oriented architecture. In: Proceeding the 5th IIAI International Congress on Advanced Applied Informatics (IIAI AAI 2016), Special Session ESKM, pp. 234–239 (2016)
6. Hisazumi, K., Yamasaki, T., Fukuda, A.: Toward impact analysis for uncertain software project. In: Proceeding IEEE International Conference TENCON 2015, 2 p (2015)
7. Hirakawa, G., Hisazumi, K., Nagatsuji, R., Nakanishi, T., Fukuda, A.: A traceability tool for model-based development dealing with uncertainties. In: Proceeding the 4th International Conference on Advances in Information Processing and Communication Technology (IPCT2016) (2016)
8. Hiya, S., Hisazumi, K., Fukuda, A., Nakanishi, T.: clooca : web based tool for domain specific modeling. In: Proceeding ACM/IEEE the 16th International Conference on Model Driven Engineering Languages and Systems (MODELS 2013), 5 p (2013)
9. Ando, T., Yatsu, H., Kong, W., Hisazumi, K., Fukuda, A.: Translation rules of SysML state machine diagrams into CSP# toward formal model checking. Int. J. Web Inf. Syst. **10**(2), 151–169 (2014)
10. Yamagata, Y., Kong, W., Fukuda, A., Nguyen, V.T., Ohsaki, H., Taguchi, K.: Formal semantics of extended hierarchical state transition matrix by CSP. Formal Aspects Comput. **26**(5), 943–962 (2014)
11. Kong, W., Liu, L., Ando, T., Yatsu, H., Hisazumi, K., Fukuda, A.: Facilitating multicore bounded model checking with stateless explicit-state exploration. Comput. J. **58**(11), 2824–2840 (2015)
12. Kong, W., Liu, L., Ando, T., Yatsu, H., Hisazumi, K., Fukuda, A.: Facilitating multicore bounded model checking with stateless explicit-state exploration. Comput. J. **7**, 17 p (2014)
13. Kong, W., Hou, G., Hu, X., Ando, T., Hisazumi, K., Fukuda, A.: Garakabu2: an SMT-based bounded model checker for HSTM designs in ZIPC. J. Inf. Secur. Appl. **31**, 61–74 (2016)
14. Liu, L., Kong, W., Ando, T., Yatsu, H., Fukuda, A.: An improvement on acceleration of distributed SMT solving. In: Proceeding the Sixth International Conference on Future Computational Technologies and Applications (Future Computing 2014), pp. 69–75 (2014)
15. Ishida, S., Tomishige, K., Izumi, A., Tagashira, S., Arakawa, Y., Fukuda, A.: Implementation of on-demand indoor location-based service using Ad-Hoc wireless positioning network. In: Proceeding the 11th IEEE International Conference on Ubiquitous Intelligence and Computing (UIC-2014), pp. 34–41 (2014)
16. Tomishige, K., Ishida, S., Tagashira, S., Fukuda, A.: Toward sensor localization using WiFi-AP anchors: realtime AP-RSS monitoring using sensor nodes. In: Proceeding the 16th Annual International Workshop on Mobile Computing Systems and Applications (ACM HotMobile 2015) (2015)
17. Ishida, S., Mimura, K., Liu, S., Tagashira, S., Fukuda, A.: Design of simple vehicle counter using sidewalk microphones. ITS European Congress (2016)
18. Ishida, S., Kunihiro, Y., Izumi, K., Tagashira, S., Fukuda, A.: Design of WiFi-AP operating channel estimation scheme for sensor node. In: Proceeding 2016 Ninth International Conference on Mobile Computing and Ubiquitous Networking (ICMU 2016) (2016)

19. Sano, Y., Yamaguchi, K., Mine, T.: Automatic classification of complaint reports about city park. Inf. Eng. Express **1**(4), 119–130 (2015)
20. Wada, R., Tagashira, S., Ogino, M., Ishida, S., Fukuda. A.: A footprint matching method for walking users in privacy-aware user tracking system using pressure sensor sheets. In: Proceeding IEEE International Conference on Agents (IEEE ICA 2016), Workshop of Sustainable Smart Mobility Platform (SSMP) (2016)
21. Holzmann, G.J., Florian, M.: Model checking with bounded context switching. Formal Aspects Comput. **23**(3), 365–389 (2011)
22. Watanabe, M.: Extended Hierarchy State Transition Matrix Design Method - Version 2.0, CATS Technical Report (2011)
23. Wynne, B.: Managing and Communicating Scientific Uncertainty in Public Policy, Background Paper, Harvard University Conference on Biotechnology and Global Governance: Crisis and Opportunity (2001)
24. Bass, L., Weber, I., Zhu, L.: DevOps - A Software Architecture Perspective. Addison-Wesley, Boston, MA (2015)
25. Shang, W.: Bridging the divide between software developers and operators using logs. In: Proceeding of the 34th International Conference on Software Engineering (ICSE'12), pp. 1583–1586 (2012)

Model-Based Methodology Establishing Traceability Between Requirements, Design and Operation Information in Lifecycle-Oriented Architecture

Shota Ishibashi, Kenji Hisazumi, Tsuneo Nakanishi and Akira Fukuda

Abstract Many uncertainties arise in system developments since the system became complex, large-scale, and deployed in uncertain environment. Developers cannot determine some type of uncertainties until start of operation. In order to deal with changing requirements and environments caused by these uncertainties, we have to tackle with building an architecture that takes operations into account in a system development lifecycle. In this paper, we propose a lifecycle-oriented development process which improve the requirement and design in terms of uncertainties for realizing sustainable information architecture for smart mobility. As a case study, we apply the proposed method to a development of Patrash which is a transfer guide application.

Keywords Lifecycle-oriented architecture · Traceability · Operation information · Smart mobility

1 Introduction

To deal with uncertainties in a system development is one of a big challenge in the software engineering field. Many kinds of uncertainties in a software development occur since the system is complex, large-scale, and deployed in uncertain

S. Ishibashi (✉)
School of Engineering, Kyushu University, Kyushu, Japan
e-mail: ishibashi@f.ait.kyushu-u.ac.jp

K. Hisazumi
System LSI Research Center, Kyushu University, Kyushu, Japan
e-mail: nel@slrc.kyushu-u.ac.jp

T. Nakanishi
Faculty of Engineering Department of Electronics Engineering
and Computer Science, Fukuoka University, Fukuoka, Japan
e-mail: tun@fukuoka-u.ac.jp

A. Fukuda
Faculty of Information Science and Electrical Engineering, Kyushu University,
Kyushu, Japan
e-mail: fukuda@ait.kyushu-u.ac.jp

© Springer International Publishing AG 2018
T. Matsuo et al. (eds.), *New Trends in E-service and Smart Computing*,
Studies in Computational Intelligence 742, https://doi.org/10.1007/978-3-319-70636-8_4

environments. Some types of uncertainties cannot be determined by actual operation of the system.

We are tackling with establishing a lifecycle oriented system architecture for facilitating system design, development, and construction for smart mobility. A sustainable information infrastructure/architecture for smart mobility is proposed [1]. Smart mobility optimizes or adjusts interests between stakeholders related to transportation such as trains, taxis, autonomous cars, walking, etc. [5]. A lifecycle oriented system architecture is needed to realize smart mobility services.

The architecture should take into account not only system design and development but also operations. The establishment of feedback technology from the operation yields the flexibility to manage changes in requirements, improvement of the system and leading to new functions or services. This study focuses on the uncertainties that occur during system development in order to realize the feedback from operation information to requirements and design.

In this paper, we propose a development process which realize feedback of the information obtained in the operation of system into requirements and design. We expand the argument in order to determine the uncertainties appeared in the development and analyze the information to be collected at operation. And, we realize the improvement of requirements and design using operation information by establishing the traceability between operation information and uncertainties.

2 Related Works

2.1 Concept Arrangement of Uncertainty

Nakanishi [2] argues the definition and classification of uncertainty. Uncertainty is defined as "necessary information in system development which is recognized by developer, defined, named, and unresolved". This definition does not regard something found after development phase as uncertainty. Moreover, uncertainty is not a vague information but need to be defined the expected result such as options or parameters.

Uncertainty is classified in the kind of decisions as follows.
Options Type

Options type uncertainty is the decision to choice someone as solution in the solution candidates.

Parameter Type

Parameter type uncertainty is the decision to determine some parameters such number or string.

Mixture Type

> Mixture type uncertainty is the abstract uncertainty which is divided to low-level uncertainties more concrete options type, parameter type and mixture type.

> We deal with uncertainty by this definition in this paper.

2.2 Requirement, Deployment, and Design Model Including Uncertainty [3]

An building method of requirements, deployment and design model for the system of which specification can not be decided at the early part of development and is to be gradually decided with the development progress is proposed in this literature. The following is outline of the model construction method.

Step 1: Making an Uncertainty Sheet

> Uncertainty sheet is a list of uncertainties to manage to all of them and enable us to trace all artifacts developed in the process. The sheet includes title, classification, impact area, etc. of uncertainty.

Step 2: Making a Case Sheet

> This case sheet is derived from the uncertainty sheet made at step 1. The step 2 we analyze changeable situations, the impact area, and incidence rate for each uncertainty.

Step 3: Making a Feature Model

> We perform feature-oriented approach based on requirements specifications and make a feature model. In order to adapt to show uncertainties, the certain element as mandatory feature and the uncertain element as optional feature are described in the model diagram. In this way, the uncertainties of the system inside are clarified.

Step 4: Making Partial Model Including Uncertainties

> The case non-common parts which can occur in different situations are exist, we make a model diagram which shows to realize all situations. This model diagram is called partial model.

Step 5: Revising Models after Determined Uncertainty

> The determined uncertainty and the point influenced by the uncertainty in the model, we can revise the point easily.

This method, by describing all revision patterns as partial model, reduce development rework cost when the uncertainty is determined. However, it is said that increase of workload by making partial model is disadvantage.

Though Sect. 2.2 method is different in considering operation step, it is excellent to know in analyzing uncertainty by making uncertainty sheet and localizing influences.

2.3   GSN(Goal Structuring Notation)

GSN is a notation of goal-oriented analysis that shows how goals are broken down into sub-goals until a point which can be supported by direct reference to evidence [4]. The principal symbols of GSN are the five as follows.

- Goal: the claim to be argue about the system
- Strategy: the approach to divide a goal to some more detailed sub-goals
- Solution: the evidence to support goals directly such specifications
- Context: the assumption to argue about goals
- Undeveloped Goal: the goal which is not able to be expanded because evidences are not exist

An example of argument using GSN is shown in Fig. 1. First, we define a top-goal such as "system is safe" and perform goal-oriented analysis. Then, the goal is applied Strategy which is described as the method to divide the goal and divided into three substantiated goals. Finally, if the goal which can be supported by test results or reports is exists, these are adopted to the goal as Solution. In case that Solution can not be adopted, the goal is set as Undeveloped Goal and is not expanded.

2.4   D-Case

D-Case is an extended notation of GSN and enable to show argument considering overall lifecycle of system development including operation. D-Case is added a new concept "Monitor" to GSN. To describe the content to be checked at operation in the Monitor symbol can be used like Solution.

The proposed method perform goal-oriented analysis using D-Case in order to show the argument about uncertainty to analyze operation information.

2.5   Project Overview of Smart Mobility Platform

Fukuda et al. [1] proposes a sustainable information infrastructure for smart mobility. The project pursues establishing an information infrastructure architecture and a seamless development methods chain for it. The project has mainly two features: (1) applying life cycle oriented methods, which are a cycle from system development to operations, to the real world, and (2) dealing with uncertainty when a system begins to be designed.

In this paper, we propose a development process to deal with uncertainty of requirements and design of the system according to this smart mobility platform.

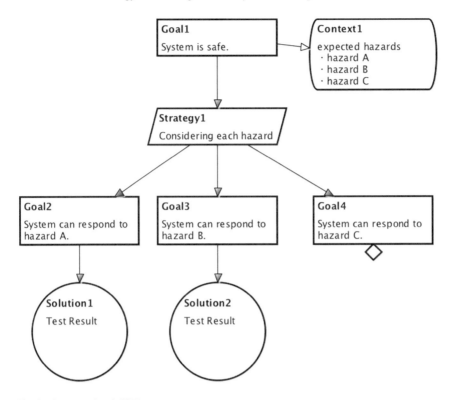

Fig. 1 An example of GSN

2.6 Operation Phase Metrics for Smart Mobility Platform

Hisazumi et al. [5] analyzed requirements of the platform for smart mobility. First, they defined smart mobility as a service to optimize or adjust interests between stakeholders related to the transportation of both of human being and things. And they conducted I* modeling and GQM.

1. Requirement analysis using I* framework, which facilitates in identifying the goals between stakeholders.
2. Elicit metrics to evaluate the goals by conducting Goal-Question-Metric (GQM).

   They proposed a methodology to identify metrics from requirements and demonstrate it for smart mobility platform. However, they do not consider the traceability between identified metrics and models of requirements and designs. In this paper, we propose a method to elicit metrics which has traceability between metrics and the model of requirement and design of the system and a development process.

3   Proposed Method

We propose a development process in order to feedback from operation information into the requirements and design. In this method, we analyze the information to be collected at operation in accordance with the uncertainties appeared in the requirements and design and establish traceability between operation information and uncertainty. After obtaining the operation information, improvement of requirements and design using operation information is realized by determining uncertainties and modifying the model diagrams based on the determined result.

3.1   Outline of Proposed Method

The proposed method introduces three works as follows to common system development and constructs a development cycle. First, we make uncertainty tables that summarize uncertainties found at requirements and design step. Next, we analyze operation information that what should be collected in order to determine the uncertainties in requirements and design. Finally, after obtaining operation information, we analyze the improvement plan to determine uncertainties using obtained information and to revise the requirements and design model diagrams.

3.2   Procedure

We describe the procedure of the proposed method with reference to a simple example. As an example, we consider a simplified information app that display the corresponding information by pressing the button "Clock", "Weather" and "News". Now, the system don't display any information at startup. So we want to improve the usability by determining the contents to be displayed at startup using operation information.

3.3   Requirement Analysis Between Stakeholders

In this step, we employ I* framework to analyze to define stakeholders related to the system and soft goals between stakeholders. User and service provider are defined as stakeholders in this information app example. And then, we analyze soft goals between user and service provider. In this example, "user can get important information quickly" is defined as a soft goal between user and service provider (Figs. 2 and 3).

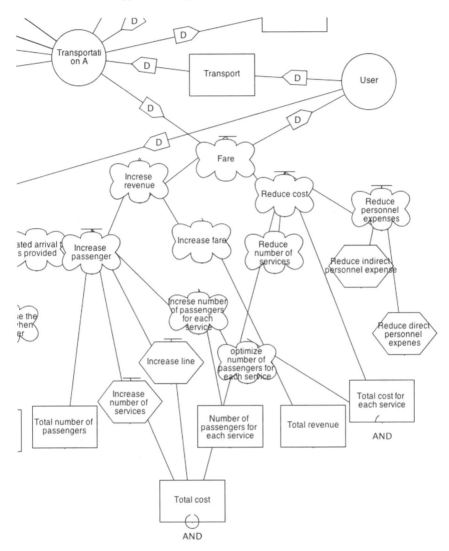

Fig. 2 A part of the model of I* and GQM [5]

3.4 Making Uncertainty Tables

First, in this step, we identify the uncertainties about the requirements and design. Uncertainties are about the requirement and design related to the leaf of the soft goals identified in the previous step. We list the uncertainty that developers cannot determine in order to achieve or improve the soft goal with reference to such requirements specification.

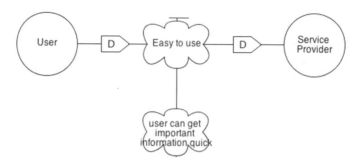

Fig. 3 The soft goal between user and service provider

Then, to each of the listed uncertainties, we describe the number of uncertainties, options of uncertainties, impact area, dependency and determined result in the table. In this paper, this is referred to as uncertainty table.

In the Options column, we describe the candidates of the determined result. The method to describe is different from uncertainty types.

- Options Type We list the all options.
- Parameter Type We describe the name of uncertain parameter. If there are some preconditions about each parameter, we also describe it.
- Mixture Type Mixture type uncertainty is the mix of options type and parameter type uncertainties. The case of mixture type, we divide the uncertainty for all options and describe these as new uncertainties. We describe the new uncertainties in the same way of parameter type.

In the column of Impact Area, the name of model diagrams which can be changed by the determined result of uncertainty is described. If dependencies between uncertainties are exist, it is described in the column of Dependency. In the column of Determined result, we describe the name of determined option or parameter, but if the uncertainty is still not determined, we set the column blank.

In this example, "the best information to be displayed at startup" is raised as uncertainty. Table 1 shows the uncertainty table of this case. The screen transition diagram is likely to be changed by the determined result of the uncertainty "the information to be displayed at startup". Then, we mark the impact area of uncertainty in the diagram. Figure 4 shows the screen diagram of this system. The point to be changed by the determined result of the uncertainty "the information to be displayed at startup" is marked as uncertainty number 1.

Table 1 Uncertainty table of simple information app

	Uncertainty name	Options	Impact area	Dependance	Determined result
1	The information displayed at startup	Watch weather news	Screen transition diagram	None	

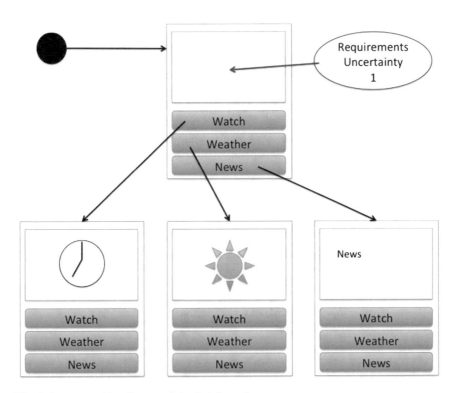

Fig. 4 Screen transition diagram of simple information app

3.5 Operation Information Analysis

3.5.1 Creating an Uncertainty Decision Making Model

In this step, we expand argument of the order to determine uncertainties in accordance with the uncertainty decision making model created at the previous Sect. 3.4. We perform goal-oriented analysis using GSN and D-Case introduced in Sect. 2.3, and create the argument diagram of uncertainties. In this paper, this is referred to as uncertainty model.

In an uncertainty decision making model, "Uncertainty X is optimally determined" is defined as a top goal and goal-oriented analysis is expanded. If the goals, which are divided to from a top goal, are supported by evidence during the development phase, the evidence is adopted to the goal as Solution. Otherwise, that is the case evidences do not exist until the operation phase, the operation information is adapted to the goal as Monitor to determine uncertainties. These methods follows the notation D-Case, which is an extension of GSN. After goal-oriented analysis about all uncertainties described in the uncertainty table, the information put as Monitor is the information to be collected at operation. If new options for the uncertainty occur by goal-oriented analysis, new options are added to the uncertainty table.

Figure 5 shows an uncertainty decision making model of the simple information app as an example. This is the uncertainty model of "the information to be displayed at startup", "the information to be displayed at startup is optimally determined" is set as top-goal, and then, the top-goal is divided into 3 sub-goal by applying the Strategy "considering each option". Evidences cannot support the 3 goals "Watch", "Weather" and "News" because user's requirements are not clear at the development phase. Therefore, we describe the operation information which can be criterion to

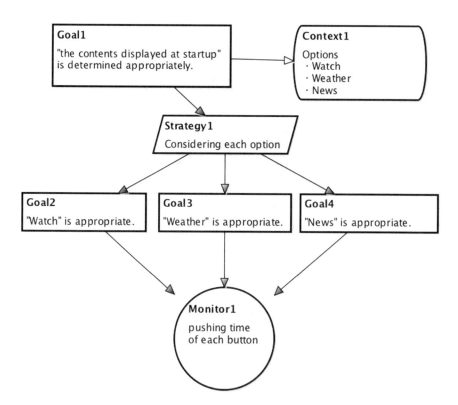

Fig. 5 An uncertainty model of simple information app

Table 2 Operation information table of simple information app

Number	Name	Corresponding uncertainty	Implemented
1	Pushing times of each button	Requirements uncertainty 1	Done

determine the uncertainty as Monitor. In this example, the Monitor "pushing times of each button" is described as the operation information to be collected. By referring to "pushing times of each button", we determine optimal one of the three "Watch", "Weather" and "News".

3.5.2 Making Operation Information Table

Next, we make a table of informations to be collected in operation described as Monitor in uncertainty models. This is referred to as operation Information table. In this table, each operation information is assigned a specific number to. Moreover, "corresponding uncertainty" and "implemented" is described. Table 2 shows the operation information table of the simple information app. By reference to Fig. 5, "pushing times of each button" is described in the table as the operation information to be collected. Corresponding uncertainty is "the information to be displayed at startup", so we describe "requirements uncertainty number 1" in the column.

3.6 Improvements Analysis

After starting operation of system and obtaining operation information, we track uncertainties from operation information and consider to determine uncertainties. In this paper, this step is referred to as Improvements Analysis. If uncertainties are determined, we revise the corresponding point in requirements and design models. This realizes feedback from operation into requirements and design.

The method of tracking from an operation information to the point of models to be revised is as follows. First, we check the uncertainty number corresponding to the operation information by referring to the table of operation information. In this example, the uncertainty corresponding to "pushing times of each button" is "the information to be displayed at startup". Next, we determine the uncertainty and describe the determined result in the column of "determined result". Finally, we check model diagrams described as impact area in uncertainty table and revise the point where the corresponding number is marked in the diagram.

Determining uncertainties using information obtained from actual operations can be viewed that the soft goal which is the parent of uncertainty is achieved or improved. By these steps, the improvement of the requirements and design of the system using operation information.

4 Case Study

In order to verify the validity of the proposed method, we conducted a case study. We applied the proposed development process to the development of real system. The target system is the transfer guide app "Patrash".

4.1 About Patrash

Nakamura et al. [6] presents personalized transportation recommendation system and implemented an Adaptive User Interface (AUI) agent called Patrash. Patrash is a transfer guide app for the porpoise of reducing the input by users as mush as possible by the information recommendation as one of the smart mobility application [7]. One of the features of Patrash is 3D interface. Therefore there are many uncertainties for developers about how it is possible to reduce inputs. There are specifications which can not be clearly determined during development phase, so it is impossible to determine uncertainties until a start of operation.

4.2 Executed Procedure

Applying proposed development process to the system divides the development into seven phases requirements analysis, design, operation information analysis, implementation, operation and improvements analysis.

1. Requirements Analysis Phase
   In this phase, we created requirements model diagrams, in addition, requirements uncertainty table. A use case diagram, a screen transition diagram, and requirements uncertainty table was created by pumping developers for information and analyzing the source code. At first, we identified stakeholders related to this system according to I* framework and "user" and "service provider" are defined as stakeholders of this system. Then, we defined "user can get optimal information when he use app" as soft goal between user and service provider. Figure 6 shows the model according to I* framework. As the result that we identified the uncertainties related to the leaf of soft goal "user can get optimal information when he use app", four uncertainties "displayed screen at startup", "displayed contents

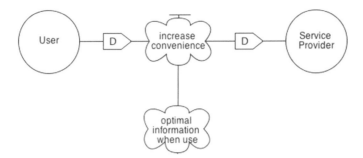

Fig. 6 The requirement model according to I* framework

Table 3 Uncertain table of Patrash

	Uncertainty name	Options	Impact area	Dependance	Result
1	Displayed screen at startup	Bus recommend screen bus search screen	Use case diagram screen transition diagram	None	
2	The total number of screens	4	Screen transition diagram	None	
		6			
3	......				
4	Necessity to display search history	Necessary unnecessary	Use case diagram screen transition diagram	None	

of each screen", "the total number of screens" and "necessity of the function to display search history" which were found as uncertainty about requirements were described in the uncertainty table. Table 3 shows a created uncertainty table. Figure 7 shows a created screen transition diagram.

2. Design Phase

In the design phase, a class diagram and a sequence diagram were created by analyzing the source code. We did not make a design uncertainty table because uncertainty about design was not found.

3. Operation Information Analysis Phase

In the operation information analysis phase, we made uncertainty models and an operation information table. Uncertainty models about four uncertainties described in the uncertainty table made at requirements analysis phase. Figure 8 shows an uncertainty model about the requirements uncertainty "displayed screen at startup". The uncertainty table shows that the options of uncertainty "the screen at startup" are "bus recommendation screen" and "bus search screen". Because the degree of use can be judged by comparing the viewing time and screen used more often should be displayed at startup, we put "viewing time of each screen" as Monitor to the goals. After creating uncertainty models about

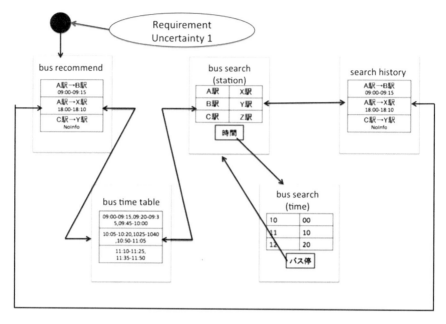

Fig. 7 Screen transition diagram of Patrash

four uncertainties, we listed each information described as Monitor in the operation information table. Table 4 shows the created operation information table.

4. Implementation Phase
   In this case study, it is decided that only "viewing time of each screen" of the
   three operation information is to be collected. And then the function to collect
   the operation information "viewing time of each screen" was implemented by
   the developers.

5. Operation Phase
   After release of the system, operation phase started. We collected operation
   informations by releasing the system to students and getting them to use.

6. Improvements Analysis Phase
   As improvements analysis, we consider the determination of the uncertainty "the
   screen at startup" using operation information "viewing time of each screen"
   and modify models. We calculated the percentage of the viewing time, of bus
   recommendation screen and bus search screen, in total time to the end after
   starting the app. Collected use history was 73 times, In these, the case that
   viewing time of "bus recommendation" is more than that of "bus search screen"
   was 52 times. Because "bus recommendation" was more used than "bus search
   screen", we judged that "bus recommendation screen" is suitable for "the screen
   at startup". Therefore we determined the uncertainty "the screen at startup"
   and started revising models. First of revising model, the column of "determine
   result" in the uncertainty table was modified from blank to "bus recommendation

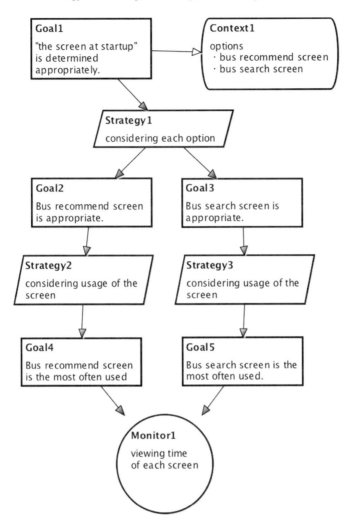

Fig. 8 Uncertainty model of "the screen at startup"

screen". Next, the point marked as "requirements uncertainty number 1" in the screen transition diagram was modified with reference to the uncertainty table.

4.3 Evaluation

We determined an uncertainty in requirements using operation information and revise the requirements model diagram. Consequently, the soft goal "user can get opti-

Table 4 Operation information table of Patrash

Number	Operation information name	Corresponding uncertainty	Implemented
1	Viewing time of each screen	Requirements 1, 2	Done
2	Screen transition times	Requirements 2	Not yet
3	Times the search history was displayed	Requirements 4	Not yet

mal information when he use app" got better and requirement of user to service provider was improved. As a result, the traceability between operation information and requirements model diagram was confirmed. Issues which have been raised are as follows.

In this case study, uncertainty at the design phase was not found. We need to apply and validate this method to the uncertainty in design and the case that uncertainty in requirements influences design. In addition, during marking uncertainty number in model diagram, if the impact area of uncertainty is large, the readability of the point to be revised in model diagrams gets worse or the point becomes ambiguous.

Furthermore, because of creating uncertainty models about all uncertainties found, the larger systems become, the more workload increases. Reducing workload by the method, such as analyzing incidence or importance about each uncertainty and creating uncertainty models about only low-importance uncertainties, is necessary.

5 Conclusion

As a part of realizing a smart mobility platform, a lifecycle oriented architecture is needed. In this paper, we proposed a development process to realize the feedback of information obtained in the operation of the system into requirements and design. Then, we conducted a case study that we apply the proposed development process to a transfer guide app Patrash. As a result, we confirmed the traceability between operation information and requirements models and showed the validity of the proposed method.

As future work, we will apply the proposed method to the uncertainty in the design phase. And, refining the model is needed. We should analyze operation information according to requirement analysis model using I* framework for stakeholders of the smart mobility platform.

Acknowledgements This work is partially supported by JSPS KAKENHI Grant Number 15H05708.

References

1. Fukuda, A. et. al.: Towards sustainable information infrastructure platform for smart mobility-project overview. In: Proceedings of the 5th IIAI International Congress on Advanced Applied Informatics (IIAI AAI 2016), Special Session ESKM, pp. 211–214 (2016)
2. Nakanishi, T. et. al.: Dynamic SPL and derivative development with uncertainty management for DevOps. In: Proceedings of the 5th IIAI International Congress on Advanced Applied Informatics (IIAI AAI 2016), Special Session ESKM, pp. 244–249 (2016)
3. Chen, C. et. al.: Requirement, deployment, and design model including uncertainty. In: Proceedings of IPSJ Embedded System Symposium (ESS), pp. 75–80 (2013)
4. Tim, K., Weaver, R.: The goal structuring notation—a safety argument notation. In: Proceedings of the Dependable Systems and Networks 2004 Workshop on Assurance Cases (2004)
5. Hisazumi, K. et. al: Operation phase metrics for smart mobility platform. In: IEEE International Conference on Agents (ICA), pp. 150–153. IEEE
6. Nakamura, H. et al.: Adaptive user interface for personalized transportation guidance system. Tourism Informatics, Selected Papers Publication in AAI 2014. Springer, pp. 119–134 (2015)
7. Patrash developer Party. Patrash3D. https://play.google.com/store/apps/details?id=com.Company.Patrash (Accessed 2016-02)

Sports Game Summarization Based on Sub-events and Game-Changing Phrases

Yuuki Tagawa and Kazutaka Shimada

Abstract Microblogs are one of the most important resources for natural language processing. This paper describes a summarization task of sports events on Twitter. We focus on an abstractive approach based on sub-events in the sports event. Abstractive summaries usually are better than summaries generated by extractive approaches in terms of readability. Furthermore, our method can incorporate sophisticated phrases that explain the scene. First, our method detects burst situations in which many users post tweets when a sub-event in a game occurs. Tweets in the burst situations are the inputs of our method. Next, it extracts sub-event elements (SEEs) that contain actions in a game, such as "Player A made a pass to Player B" and "Player B made a shot on goal." Then, it identifies the optimal order of the extracted SEEs by using a scoring method. Finally, it generates an abstractive summary on the basis of the ordered SEEs, such as "Playler B made a shot on goal from the Player A's pass." In addition, it adds game-changing phrases into the abstractive summary by some rules. In the experiment, we show the effectiveness of our method as compared with related work based on an extractive approach.

1 Introduction

The World Wide Web contains a huge number of online documents that are easily accessible. Analysis of the documents has an important role for natural language processing. On the web, microblog services have become more important in recent years as resources for NLP. Twitter is one of the most famous microblogging services

Y. Tagawa (✉)
Graduate School of Computer Science and Systems Engineering,
Kyushu Institute of Technology, 680-4 Kawazu Iizuka,
Fukuoka 820-8502, Japan
e-mail: y_tagawa@pluto.ai.kyutech.ac.jp

K. Shimada
Department of Artificial Intelligence, Kyushu Institute of Technology,
680-4 Kawazu Iizuka, Fukuoka 820-8502, Japan
e-mail: shimada@pluto.ai.kyutech.ac.jp

© Springer International Publishing AG 2018   65
T. Matsuo et al. (eds.), *New Trends in E-service and Smart Computing*,
Studies in Computational Intelligence 742, https://doi.org/10.1007/978-3-319-70636-8_5

and text-based posts of up to 140 characters. The posted sentences are referred to as "tweets." Tweets tend to be posted as lifelog data in real time. One of the most famous Twitter-broadcasts, live-tweet, is play-by-play comments of sports events such as soccer and baseball games. Summarizing tweets related to a sports event is one of the most interesting researches of practical use.

Nichols et al. [6] have proposed a summarization method for sports events using Twitter. Kubo et al. [3] have proposed a method generating sports updates that leverages good reporters on Twitter. These studies focused on a phenomenon in which many users post tweets. It is called a burst [2]. They generated a summary of an event by automatically extracting important tweets in each burst. However, extractive summarization has some problems in terms of redundancy of the generated summary. Summaries based on extractive methods often contain not only suitable information for the summary but also private opinions, emotions and figurative expressions.

ex1.   Yeeeeeeeees!!! Got it!!! Messi, the genius, scored a goal with Suarez's pass when I took my eyes off of TV just for a second.

This tweet is an example of an output from an extractive method. It contains many problems as a summary. "Yeeeeeeeees!!!" is not informative because it is an emotional phrase. "when I took my eyes off of TV just for a second" is just the user's situation. "the genius" is also not required in the summary.

To solve these problems, we propose an abstractive summarization method using tweets. In our method, we also handle burst situations for the 1st step of the summarization. Next, our method extracts sub-event elements (SEEs), which are inputs for the summarization, from each burst in an event. Then, it estimates the importance of each SEE and the optimal order among them on the basis of a scoring function. Finally, it generates an abstractive summary from the ordered SEEs.

ex2.   The opening goal of Barcelona.
ex3.   Messi scored a goal with Suarez's pass. This is the opening goal of Barcelona.

In this example, assume that ex2 is also an important tweet in a burst related to ex1. The purpose of our method is to generate a summary ex3 from ex1 and ex2.

The ex3 is a better summary as compared with the ex1. However, we can not capture the situation of the sub-event from only the ex3. Therefore, our method incorporates sophisticated phrases that explain the scene by using some rules. We call them game-changing phrases (GP). For example, assume that the goal in the ex3 occurs in the 10 min. In this situation, our method adds the GP, "at an early stage", into the ex3.

ex4.   Messi scored a goal with Suarez's pass *at an early stage of the game*. This is the opening goal of Barcelona.

If the goal occurs in the 80 min, namely later in the game, our method adds the GP, "tie-breaking", into the ex3.

ex5.   Messi scored a *tie-breaking* goal with Suarez's pass. This is the opening goal of Barcelona.

These GPs help readers instantly understand the situation.

The contributions of this paper are as follows:

- We generate an abstractive summary from Twitter on the basis of sub-events in a sports game.
- We incorporate the abstractive summary with a game-changing phrase.
- We compare the proposed method with an extractive approach from some points of view.

2 Related Work

There are many extractive summarization methods for microblogs [3, 6, 10]. Takamura and Okumura have proposed a summarization model for a stream, tweets, on Twitter. They formulated the summarization task as the facility location problem [9]. Nichols et al. [6] extracted important tweets by using a sentence ranking method based on phrase graphs. However, the method tended to extract emotional tweets such as screaming and applause at the beginning of sub-events such as goals. Kubo's method improved this problem by introducing new factors, namely specific words, such as shoot and yellow card in the soccer domain, and good reporters. One advantage of the extractive summarization methods is that the methods can relatively generate grammatically well-formed summaries because the methods just extract tweets on Twitter. On the other hand, abstractive summarization has the important role to generate more informative summaries without redundant elements.

As an abstractive summarization method, Sharifi et al. [7] have reported an automated summarizer based on the phrase reinforcement algorithm for Twitter. They generated a word graph, and then detected the most heavily weighted path as the final summary. Although the method was flexible, the target of the summarization was not a stream, namely tweets on timeline. In this paper, we propose an abstractive summarization method on timeline.

3 Proposed Method

The proposed method consists of two parts; (1) burst detection and (2) summarization.

3.1 Burst Detection

The burst in this paper is a point in which the number of tweets suddenly increases. In other words, it indicates that an important sub-event in a game has occurred; e.g., goals in soccer games.

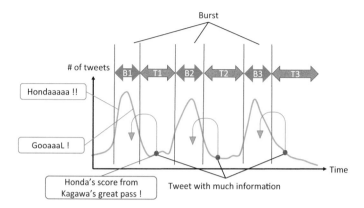

Fig. 1 The bursts on timeline

Our method calculates the median value of the number of tweets per 30 seconds (*AllMed*30). If the number of tweets in a thirty-second period is more than *AllMed*30 × 3, we regard the period as a burst. Figure 1 shows an example of burst situations. In the figure, B1, B2 and B3 are bursts respectively.

Here we introduce a rule based on heuristics. Tweets in a burst often tend to be short messages because users are enthused about the situation, such as someone's goal; "Hondaaaaa!!" and "GooaaaL!" On the other hand, tweets after a burst often contain explanatory words or phrases; "Honda's score from Kagawa's great pass!" Therefore, we need to add such tweets into the input of the next step, summarization. For the addition, we extract player's names and specific words, such as "goal", in the burst first. Then, tweets containing the player's names and specific words are extracted from tweets right after the burst.

3.2 Summarization

The input of this summarization process is tweets in each burst. The summarization process consists of three processes; (1) sub-event element (SEE) extraction, (2) sub-event element (SEE) ordering and selection, and (3) summary generation. As an optional function, our method integrates a summary with game-changing phrases. Figure 2 shows the outline of the summarization process.[1]

---

[1] Note that our target is Japanese tweets although the examples are written in English.

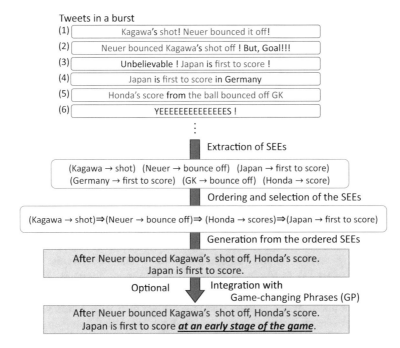

Fig. 2 The outline of the summary generation

3.2.1 Sub-event Element Extraction

First, our method extracts sub-event elements (SEEs) from tweets in each burst. An SEE is a relation between an agent and an action, and is expressed by the symbol $\rightarrow$. It is based on the dependency relation between an agent and an action in a tweet. More precisely, the symbol $\rightarrow$ denotes a dependency from a noun except specific words and verbal nouns to a verb, a verbal noun or a specific word. In Fig. 2, the SEEs (Kagawa $\rightarrow$ shot) and (Neuer $\rightarrow$ bounce off) are extracted from the tweet (2) "Neuer bounced Kagawa's shot off! But, Goal!!!"

For the SEE extraction, we use Cabocha, a Japanese dependency analysis tool [5]. In general, tweets contain grammatical errors, such as omission of particles, and Internet slangs. The errors often cause mistakes of dependency analysis [1]. To solve this problem, our method estimates an appropriate particle and adds it into each tweet in advance if there is omission of a particle.[2] We use Google Web Japanese n-gram [4] and some hand coded rules for the estimation. Then, our method extracts SEEs by using Cabocha. In addition, we apply the transitive law to the extracted SEEs. In other words, we generate (a $\rightarrow$ c) if there are (a $\rightarrow$ b) and (b $\rightarrow$ c).

---

[2]The target of this process is only some combinations of words. For example, "a team name + a specific word"; Japan goal (日本先制) should be Japan's goal (日本が先制).

3.2.2 Sub-event Element Ordering and Selection

We obtain many SEEs in Sect. 3.2.1. Moreover, we need to estimate ordering of the SEE for the generation process. Therefore, we introduce two processes, ordering and selection of SEEs, for the summary generation in the next section. The basic policies are as follows:

- If an SEE see_1 is more frequent than an SEE see_2, see_1 is more important than see_2.
- If an SEE see_2 contains the same word in an SEE see_1 with high frequency, see_2 is rejected because it is redundant.

First, we estimate the appropriate sequence of SEEs. The order is expressed by the symbol $\Rightarrow$. If two SEEs are expressed by $(x1 \rightarrow y1) \Rightarrow (x2 \rightarrow y2)$, it indicates that $(x2 \rightarrow y2)$ follows $(x1 \rightarrow y1)$. For the estimation, we use the dependency relation between words. For example, we can estimate that the order of the SEEs, (Kagawa $\rightarrow$ shot) and (Neuer $\rightarrow$ bounce off), is (Kagawa $\rightarrow$ shot) $\Rightarrow$ (Neuer $\rightarrow$ bounce off) from the tweet (1) in Fig. 2. In other words, the word "bounce off" tends to frequently appear after the word "shot" in this situation because the action "bounce off" is brought about by the action "shot".[3] Here we deal with words with similar meanings. For example, the word "over-the-head shot" fits into the category of "shot". We obtain the category word lists by using a hierarchic structure in each entry in Wikipedia. For the word "shot", we can acquire 12 words, such as a middle-range shot, a long-range shot, a lob over shot, a drive shot, a header and a volley from Japanese Wikipedia.

Let me explain with an example. Assume that there are three SEEs, A, B and C, and the order of the A and B has been already determined; A(a1 $\rightarrow$ a2) $\Rightarrow$ B(b1 $\rightarrow$ b2). In other words, the frequency of the SEE C is lower than the frequencies of A and B, and b1 and b2 tend to frequently appear after a1 and a2 in tweets of the burst. In this case, there are three combinations for ordering. Figure 3 shows the combinations. Our method counts the frequencies of these combinations in tweets, and then determines the candidate location of the SEE C, namely the combination with the maximum frequency.

There are many SEEs in a burst. Not all SEEs extracted are suitable for the summary. Therefore we apply a scoring function S to the SEE selection. By using this function, we determine whether the combination with the maximum frequency is suitable or not.

$$S = \frac{maxFreq \times seeScore \times seeFreq}{wordNum \times tweetNum \times seeBurstNum} \tag{1}$$

where *maxFreq* is the maximum frequency computed for the determination of the appropriate location of an SEE. *seeScore* is a weight of the SEE and as follows:

---

[3]In addition, we also use some rules. For example, the SEE "goal" follows the SEE "assist".

$$C(c1 \rightarrow c2) \Rightarrow A(a1 \rightarrow a2) \Rightarrow B(b1 \rightarrow b2) \quad (1)$$

$$A(a1 \rightarrow a2) \Rightarrow C(c1 \rightarrow c2) \Rightarrow B(b1 \rightarrow b2) \quad (2)$$

$$A(a1 \rightarrow a2) \Rightarrow B(b1 \rightarrow b2) \Rightarrow C(c1 \rightarrow c2) \quad (3)$$

$$seeScore = \begin{cases} 4 & \textit{if} \quad \text{the SEE contains} \\ & \quad \text{(PN or TN) and SW} \\ 3 & \textit{elsif} \quad \text{the SEE contains PN} \\ 2 & \textit{elsif} \quad \text{the SEE contains SW} \\ 1 & \textit{otherwise} \end{cases}$$

where PN, TN and SW are a player's name, a team name and a specific word, respectively. *seeFreq* is the frequency of the SEE in the burst. *wordNum* and *tweetNum* are the number of words in the SEEs and the number of tweets in the burst. *seeBurstNum* is the number of bursts containing the target SEE before the current burst. If the S is more than a threshold θ, we add the SEE into the SEE list as the input for the summary generation process described in the next section.

Each SEE consists of two words, an agent and an action. As a result, generated sentences from only simple SEEs are often not explanatory. Therefore, we add complementary information into SEEs for generating a more informative summary. Here assume that the following SEEs are extracted from a sub-event.[4]

- (PlayerA $\rightarrow$ make_a_goal) : 7
- (PlayerA $\rightarrow$ PK) : 6

From the frequency, (PlayerA $\rightarrow$ make_a_goal) is selected. In addition, assume that there are some paths between PlayerA and make_a_goal in tweets in the sub-event.[5]

- <PK $\rightarrow$ make_a_goal> : 6
- <PlayerA $\rightarrow$ steadily> : 2
- <Steadily $\rightarrow$ make_a_goal> : 2

From the SEEs and paths, we can generate a graph structure as shown in Fig. 4. We compute the weight of each path in the graph as follows:

$$S_{path} = \begin{cases} 0 & \textit{if} \quad \text{the inner path contains PN or TN} \\ \frac{sumDep}{numWords} & \textit{otherwise} \end{cases} \quad (2)$$

where *sumDep* and *numWords* denote the sum of the frequencies of the dependencies and the number of words in the path, respectively. In Fig. 4, the path

---

[4]The numbers are the frequencies of each SEE.

[5]Note that these items with "<" and ">" are not SEEs. They are just dependency relations between words in tweets.

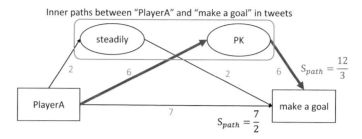

Fig. 4 Determination of the inner path for generating a more informative summary

(PlayerA $\rightarrow$ PK $\rightarrow$ make_a_goal) obtain the maximum $S_{path} = \frac{6+6}{3}$; (PlayerA $\rightarrow$ PK): 6 and <PK $\rightarrow$ make_a_goal>: 6 with three words, namely "PlayerA", "PK" and "make_a_goal".

3.2.3 Basic Summary Generation

We generate a summary from the selected and ordered SEEs, such as "After Neuer bounced Kagawa's shot off, Honda's score. Japan is first to score." in Fig. 2. The summary generation consists of two parts; determination of inflected forms of verbs and addition of particles.

The Japanese language has some inflected forms about verbs. Verbs except the last SEEs are changed to the continuative form. For example, "決める (score a shot)" is changed to "決め" and "弾く (bounce off)" is changed to "弾き". We also decide whether the voice should be active or passive on the basis of the statistics in tweets of the burst.

Next, we add particles and punctuation to the summary for increasing the readability. In other words, we replace $\rightarrow$ and $\Rightarrow$ in SEEs with particles and punctuation by using some rules described in Table 1. w1 and w2 in the table are applied into two situations; (w1 $\rightarrow$ w2) with the relation "$\rightarrow$" and (w0 $\rightarrow$ w1) $\Rightarrow$ (w2 $\rightarrow$ w3) with the relation "$\Rightarrow$."

If the SEEs do not fit into the rules, our method add particles and punctuation for generating a summary on the basis of the following rules.

Table 1 The rules for the addition of particles

Condition	Particle
w1 is a player's name or a team name and w2 is a specific word such as "score"	w1 の w2
For the last SEE, w1 is a player's name or a team name and w2 is a specific word such as "score"	w1 が w2
w1 is "クロス (cross)", "パス (pass)", "アシスト (assist)" or "キック (kick)"	w1 から w2

Table 2 Examples of rules for GPs

GP	Time condition (min)	Result condition
Dramatic equalizer	After 80	Tied score
Return to the starting point	After 50	Tied score
At an early stage	Before 10	Opening goal
Tie-breaking	After 50	Opening goal

Step1   Select the particle that follows w1 and has the maximum frequency about the dependency relation between w1 and the particle in tweets of the burst.

Step2   Select the particle that has the maximum frequency about the combination "w1+Particle+w2" in tweets of the burst.

Step3   Select the particle that has the maximum frequency about the combination "w1+Particle" in tweets of the burst. If there is no candidate, replace → or ⇒ with the punctuation (comma).

3.2.4   Incorporation with Game-Changing Phrase

A summary generated in Sect. 3.2.3 contains informative content to understand each sub-event such as a goal. However, we can not capture the situation of the sub-event from only the summary. Therefore, we generate a more informative summary by incorporating sophisticated phrases that explain the scene. We have proposed a method to acquire sophisticated phrases for baseball game's summary generation [8]. We call them Game-changing Phrases (GP). We convert the GPs for the baseball task to this situation, the soccer task.

The GPs consist of three types of phrases: (1) Time Phrases (TP),[6] (2) Action Phrases (AP) and (3) Result Phrases (RP). TPs are phrases to explain time information of the scene, e.g., "at an early stage". TPs are usually added into the beginning of the summary on the basis of the grammatical rules of Japanese. APs are phrases to modify actions, such as a goal, in the game, e.g., "worthful". APs are usually added as a modifier of the action. RPs are phrases to explain the result of the sub-event, e.g., "adding another goal". RPs are usually added into the end of the summary on the basis of the grammatical rules of Japanese.

Table 2 shows examples of GPs and rules to apply them to the summary. For example, if a team A is one point behind, the current time on the game is more than 80 min and a player X of the team A make a score, namely a tied score, then our method adds the "dramatic equalizer" into the summary as an AP; "Player X made a *dramatic equalizer$_{AP}$* score."

---

[6]In the baseball task, Inning Phrases (IP).

Table 3 The dataset

Hashtags	Teams	Date	# of tweets
#CL, etc.	Barcelona versus Juventus	2015/6/6	9945
#CWC, etc.	Barcelona versus River Plate	2015/12/20	5258
#nadeshiko, #WCup, etc.	Japan versus Nederland	2015/6/24	4734
	Japan versus Australia	2015/6/28	6831
	Japan versus England	2015/7/2	7862
	Japan versus USA	2015/7/6	16,507

4 Experiment

4.1 Dataset and Settings

We used tweets about six soccer games for the evaluation. We collected tweets with hashtags related to the target games. Table 3 shows the statistics of tweets. We also manually collected player's names and specific words from Wikipedia,[7] a soccer news site[8] and the web site of Japan Football Association[9] in advance.

We compared our method with an extractive method based on Kubo et al. [3]. Our method needs a threshold θ for the score S. For the evaluation of a game, we determined the value of θ by using other five games.

4.2 Evaluation

The evaluation is grouped into verification of our method itself, a comparison with related work and the effectiveness of integration of GPs. The criterion about the evaluation of our method was readability of the output from our method. The criteria about the comparison were a preference measure and a compression ratio among our methods and an extractive method.

---

[7]https://ja.wikipedia.org/wiki/ サッカー用語一覧 .

[8]http://web.gekisaka.jp/.

[9]http://www.jfa.jp/.

Table 4 The evaluation of the readability

Teams	Readability
Barcelona versus Juventus	3.57
Barcelona versus River Plate	3.76
Japan versus Nederland	3.14
Japan versus Australia	3.85
Japan versus England	3.33
Japan versus USA	3.88
Average	3.59

4.2.1 Readability

In general, extractive summarization methods is better in terms of generating grammatically well-formed summaries than abstractive summarization methods because the extractive methods just extract sentences from the target documents. On the other hand, grammatical accuracy, including easy-to-read summaries, is one of the most important factors in the evaluation for abstractive methods. Therefore, we evaluated our method with a 4-grade evaluation value[10] by three test subjects who were students and did not relate to this study. The three test subjects received summaries, and then evaluated them subjectively in terms of readability.[11]

Table 4 shows the experimental result. The values in the table are the average scores of the three test subjects for each game. Our method obtained high scores, namely over 3, in all games. The average score was 3.59 in the 4-grade evaluation value. This result shows that the generated sentences from SEEs are appropriate as a summary.

4.2.2 Comparison with Related Work

Next, we compared our method with an extractive method. Six test subjects who were students and did not relate to this study judged which method is suitable as a summary of the game. The six test subjects received two summaries, namely a summary by our method and a summary by Kubo's method. This evaluation was a 4-grade preference value between our abstractive method and an extractive method based on Kubo's method. Here the value 1 denotes the $method_A$ was better and the value 4 denotes that the $method_B$ was better. Note that the test subjects didn't know which document was generated by our method. In this experiment, the $method_A$ is Kubo's method and the $method_B$ is our method. Therefore, if the value is more than 2.5, our method is more suitable than the extractive method, as a summary.

---

[10] 1: low readability to 4: high readability.

[11] In the experiment, no definition about the readability was given to the test subjects. The questionnaire just contained the sentence "Evaluate the readability of each document."

Table 5 The comparison of the preference and the compression ratio (CR)

Teams	Preference	CR
Barcelona versus Juventus	3.00	58.08
Barcelona versus River Plate	3.42	51.35
Japan versus Nederland	3.14	23.26
Japan versus Australia	2.23	23.55
Japan versus England	2.78	31.19
Japan versus USA	2.38	23.51
Average	2.83	35.16

Another criterion in this section is a compression ratio between the two methods. The compression ratio is one of the most important criteria in the summarization task. In this experiment, we computed the ratio as follows:

$$CR = \frac{\text{The length of a summary from our method}}{\text{The length of a summary from Kubo's method}} \times 100 \qquad (3)$$

Here the length denotes the number of characters in a summary.

Table 5 shows the result of the preference and the compression ratio. The preference and CR values in the table are the average scores of the six test subjects for each game. The average value of the preference was 2.83. This result indicates that the summaries from our method is comparatively suitable than those from the extractive methods. In addition, the average value of the CR was 35.16. In other words, our summary was preferred as compared with the extractive method despite one-third of the length of the extracted summary. These results show the effectiveness of our method.

The difference of CR about Barcelona's games and Japan's games was caused by the type of games. Users that posted tweets in the Barcelona's games are usually enthusiastic football fans. Therefore, the tweets tended to contain explanatory words or phrases. As a result, the output of our method tended to become long. On the other hand, Japan's games in this experiment were games in the FIFA Women's World Cup 2015. Many people watched the games because the World Cup was a famous event. Therefore, tweets that were posted in the games were not always explanatory. In fact, the number of explanatory SEEs about the World Cup games was lower than those about the Barcelona's games. For the World Cup games, the number of connected SEEs for the summarization process was two or three SEEs. As a result, the generated summaries for the World Cup tended to become shorter than those for the Barcelona's games. Thus, as Kubo et al. mentioned in [3], finding good reporters is an important task for the summarization, especially in general-interest events.

Next, we compared the proposed method, namely summaries in Sect. 3.2.3 and the summaries with GPs in terms of the compression ratio (CR) with the Kubo's method. Table 6 shows the result of the compression ratio. In the table, the column of

Table 6 The comparison of the compression ratio (CR) between basic summaries and summaries with GPs

Teams	without GPs	with GPs
Barcelona versus Juventus	58.08	75.08
Barcelona versus River Plate	51.35	72.06
Japan versus Nederland	23.26	32.70
Japan versus Australia	23.55	26.52
Japan versus England	31.19	36.10
Japan versus USA	23.51	42.49
Average	35.16	47.49

"without GPs" is the same as the CR in Table 5. Although the summaries with GPs were longer than those without GPs, the summaries with GPs were shorter than the Kubo's method. In other words, the method with GPs can generate more informative summaries as compared with the method without GPs and generate more compact summaries as compared with the related work. In addition, we can easily select the method, with GPs or without GPs, on demand because the method with GPs is optional. These result show the effectiveness of our two methods.

4.3 Discussion

We now discuss the outputs with some examples, in detail. Table 7 shows actual sentences generated by our method without GPs and sentences extracted by Kubo's method. The sub-event names in the table were given by authors after the summary generation. The test subjects judged that the sentences of our method for "Save", "Goal", "Kickoff" and "Substitution" were better than Kubo's method. On the other hand, for "Dribbling", the test subjects preferred the output of Kubo's method.

For the sub-event "Save", the sentence of our method was more explanatory than Kubo's method. Our method correctly combined two SEEs extracted from different tweets in the burst and then generated the sentence from the SEEs, (Suarez → shot) and (Buffon → save). Extractive methods can not extract such sentences essentially because the descriptions related to the two SEEs appeared in different tweets. In other words, extractive methods can not essentially generate a suitable summary if some important sub-events for summarization appear in different tweets. This is one of the most significant advantages of abstractive summarization, as compared with extractive summarization.

In a similar way, our method generated more explanatory sentences from four SEEs, (Stegen → bounce off), (Marchisio → back-heeled pass), (Teves → shot) and (Morata → score). Our method can generate a concise and short summary from many tweets in a burst because it is based on an abstractive approach. On the other

Table 7 Examples of the generated sentences from our method and Kubo's method

Sub-event	Method	Sentence
Save	Ours	Buffon saved on the Suarez's shot. スアレスのシュートをブッフォンがセーブ
	Kubo's	Great saving! Buffon! グレイトセーブ！ブッフォン！
Goal	Ours	Stegen bounced the Teves's shot from Marchisio's back-heeled pass off, but Morata scored. マルキージオのヒールパスからテベスのシュートを テアシュテーゲンが弾きモラタがゴール
	Kubo's	Yeees!! That back-heeled pass was good timing ← Morata got the benefit of the whole thing, haha きたぁぁーー!! あのヒールパス素晴らしすぎる タイミングだ←モラタ美味しいなー笑
KickOff	Ours	Kickoff of the first half of the game. 前半戦のキックオフ
	Kubo's	The time for kickoff of the final 16, vs. Nederland. Kawasumi with deft dribbling. Miyama with great passes. Hope the large score in Vancouver. いよいよなでしこの決勝トーナメントオランダ戦キックオフ. 切り裂け川澄，通せ宮間，気分はバンクーバーの弾幕.
Substitution	Ours	Ohno was OUT, Iwabuchi was IN. 大野の交代，岩淵が投入.
	Kubo's	The 1st substitution of Japan. Ohno was Out and Iwabuchi was IN. Hope the speedy dribbling. Come on, Iwabuchi! 日本1枚目．大野アウトで岩淵イン，岩淵，ドリブルで切り裂いたれ!!
Dribbling	Ours	Iwabuchi dribbled. 岩淵がドリブル.
	Kubo's	Mana Iwabuchi dribbled past opposing players, and then made a shot. But the shot was off target. 岩渕真奈のドリブル突破からのシュートはゴール横へ！

hand, our method generated simple sentences that clearly explained the situation for "KickOff" and "Substitution". However, the extractive method extracted opinions and emotions of the user that posted the tweet, such as "Hope the large score in Vancouver." These expressions are not always suitable as sentences in a summary of a sports game. By using an abstractive approach, our method can delete redundant phrases in tweets.

For the sub-event "Dribbling" in the table, the information on the output from our method was insufficient, as compared with the output from Kubo's method, because the output did not contain the result of the "dribbling", namely "shot (but no goal)" in this example. In this situation, our method extracted the SEE (Iwabuchi → dribble) first as the most frequent SEE in the burst. To realize more informative sentence, our method needs to add the word "dribble" into the SEE (Iwabuchi → shot). In other words, we want to generate the SEE (Iwabuchi → dribble → shot). To solve this problem, we need to integrate two SEEs with the same agent; generation of (agent1

$\rightarrow$ action1 $\rightarrow$ action2) from (agent1 $\rightarrow$ action1) and (agent1 $\rightarrow$ action2).[12] This is important future work.

Finally, we compared the results from the methods without GPs (Basic) and with GPs (w/GP). Some examples are as follows:

Basic1:   Iwabuchi scored a goal. This is the opening goal of Japan.
(岩淵のゴール．日本が先制．)

w/GP1:   Iwabuchi scored a *worthful*$_{AP}$ goal *in the 78th minute* (*scoreless minutes*)$_{TP}$.
This is the *long-awaited*$_{RP}$ opening goal of Japan.
(両軍無得点で迎えた後半 33 分 $_{TP}$，岩淵の値千金 $_{AP}$ のゴール．日本が待望の先制点を挙げる $_{RP}$．)

Basic2:   Lloyd scored a goal.
(ロイドが決める．)

w/GP2:   *In the 5th minute of the first half*$_{TP}$, Lloyd scored a goal *again*$_{AP}$ and *USA comes out ahead*$_{RP}$.
(1 点リードの前半 5 分 $_{TP}$，ロイドが再び $_{AP}$ 決め試合を優位に進める $_{RP}$．)

The proposed method with GPs generated informative summaries as compared with that without GPs. On the other hand, there were some mistakes of summaries with GPs. The readability score was 3.03 (vs. 3.59 for the method without GPs. See Table 4.) The reason why some summaries generated from the method with GPs were ungrammatical was that a burst sometimes contained several SEEs. As a result, inappropriate integration with a GP occurred in the summaries. The following is an example:

w/GP3:   Bassett's own goal, Japan moves into final and *take the lead*$_{RP}$
(バセットのオウンゴールで日本が決勝進出でリードを奪う $_{RP}$．)

This example contained two SEEs, "own-goal" and "move-into-final". The RP, "take the lead", should be integrated with the SEE, "own-goal". The improvement of the burst detection is important future work.

5   Conclusion

In this paper, we proposed an abstractive summarization method about sports events on Twitter. First, our method detected burst situations by using an approach in the previous work. Next, it extracted sub-event elements (SEEs), which are important elements for the summarization, from each burst in an event, such as (Player A $\rightarrow$ pass) and (Player B $\rightarrow$ shot) from "Player A made a pass" and "Player B made a shot on goal." Then, it estimated the importance of each SEE and the optimal order among

---

[12]This integration is not realized in the current method because of the conflict with basic policies in Sect. 3.2.2.

them on the basis of a scoring function. Finally, it generated an abstractive summary from the ordered SEEs. In addition, it added Game-changing Phrases (GPs) into the abstractive summary by some rules.

In the experiment, we evaluated the readability of the output of our method first. We obtained the high readability score in the experiment. Then, we compared our method with an extractive summarization method. Our summary was preferred as compared with the extractive method despite one-third of the length of the extracted summary. By using an abstractive approach, our method can delete redundant phrases in tweets and generate a concise summary from many tweets in a burst. These results show the effectiveness of our method. On the other hand, the information on some outputs from our method was insufficient. We also evaluated the method with GPs. The method with GPs generated more informative summaries as compared with the method without GPs and generated more compact summaries as compared with the related work. In the experiment, we evaluated the readability of the results from our method and compared our method with related work in terms of the preference and compression ratio. We need to consider another evaluation measure for more precise evaluation. It is also our future work.

References

1. Ikeda, K., Yanagihara, T., Hattori, G., Matsumoto, K., Ono, C.: Estimation and complementation approach for the analysis of the omission of postposition on colloquial style sentences. In: IPSJ SIG Techinical Report, DBS-151(39) (in Japanese), pp. 1–8 (2010)
2. Kleinberg, J.: Bursty and hierarchical structure in streams. In: Proceedings of the eighth ACM SIGKDD international conference on Knowledge discovery and data mining, pp. 91–101 (2002)
3. Kubo, M., Sasano, R., Takamura, H., Okumura, M.: Generating live sports updates from twitter by finding good reporters. In: Proceedings of the 2013 IEEE/WIC/ACM International Joint Conferences on Web Intelligence (WI) and Intelligent Agent Technologies (IAT), pp. 527–534 (2013)
4. Kudo, T., Kazawa, H.: Web Japanese n-gram version 1. In: Gengo Shigen Kyokai (GSK2007) (2007)
5. Kudo, T., Matsumoto, Y.: Japanese dependency analysis using cascaded chunking. In: Proceedings of COLING-02 proceedings of the 6th conference on Natural language learning, pp. 1–7 (2002)
6. Nichols, J., Mahmud, J., Drews, C.: Summarizing sporting events using twitter. In: Proceedings of the 2012 ACM international conference on Intelligent User Interfaces (IUI), pp. 189–198 (2012)
7. Sharifi, B., Hutton, M.A., Kalita, J.: Summarizing microblogs automatically. In: Proceedings of HLT '10 Human Language Technologies: The 2010 Annual Conference of the North American Chapter of the Association for Computational Linguistics, pp. 685–688 (2010)
8. Tagawa, Y., Shimada, K.: Inning summarization based on automatic generated templates (In Japanese). In: Proceedings of the Twenty-third Annual Meeting of the Association for Natural Language Processing, vol. B7-3 (2017)
9. Takamura, H., Okumura, M.: Text summarization model based on facility location problem. Trans. Jpn. Soc Artif. Intell. **25**, 174–182 (2010)
10. Takamura, H., Yokono, H., Okumura, M.: Summarizing a document stream. In: Proceedings of the 33rd European conference on Advances in information retrieval (ECIR), pp. 177–188 (2011)

Headline Generation with Recurrent Neural Network

Yuko Hayashi and Hidekazu Yanagimoto

Abstract Automatic headline generation is related to automatic text summarization and it is useful to solve information flood problems. This paper aims at generating a headline using a recurrent neural network which is based on a machine translation approach. Our headline generator consists of an encoder and a decoder and they are constructed with Long Short Term Memory, which is one of recurrent neural networks. The encoder constructs distributed representation from the first sentence in an article and the decoder generated headlines from the distributed representation. In our experiments, we confirmed that our proposed method could generate appropriate headlines but in some articles this method generates meaningless headlines. The results show that our proposed method is superior to another approach, statistical machine translation from the viewpoint of ROUGE, which is an evaluation score of automatic text summarization. Furthermore, we could find that using an input sentence in reverse order improves the quality of headline generation.

Keywords Natural language processing · Neural network · Headline generation

1 Introduction

Recently there are a lot of text data in the world because of the growth of the Internet. Even though you select text data with search engines, it is impossible to read all texts because many search results are related to a search query, so we have to reduce text data that we have to read without loss of information. To achieve it automatic text summarization is paid attention to by many researchers. Automatic text summarization reduces the amount of texts without loss of information and help you understand the texts.

Y. Hayashi (✉) · H. Yanagimoto
College of Sustainable Systems Science, Osaka Prefecture University,
1-1, Gakuen-cho, Naka-ku, Sakai, Osaka 599-8531, Japan
e-mail: swa01232@edu.osakafu-u.ac.jp

H. Yanagimoto
e-mail: hidekazu@kis.osakafu-u.ac.jp

© Springer International Publishing AG 2018   81
T. Matsuo et al. (eds.), *New Trends in E-service and Smart Computing*,
Studies in Computational Intelligence 742, https://doi.org/10.1007/978-3-319-70636-8_6

Automatic text summarization has two approaches; an extractive approach and an abstractive approach. The extractive approach is a method that extracts important words or important sentences from texts according to their importance. However, the extractive approach has some restrictions. First, generated abstract consists of sentences in an original text and the sentence is not made be short even if it includes redundant expressions. Second, when the extractive approach extracts words from texts, it is difficult to construct a natural sentence from the extracted words. On the other hand, the abstractive approach is a method that generates new sentences from original texts like human activities. The approach can generate a favorable sentence but needs more natural language processing techniques to construct a sentence and it includes unsolved problems. However, in these days, the automatic generative text summarization is realized with a neural network inspired from machine translation [19].

In this paper we proposed headline generation from a body of articles with a neural network because we regard headline generation as one of automatic text summarization tasks. Usually you can imagine a context of the article by reading a headline of an article and decide whether you can read it or not. In this process a headline acts as a summary of an article, so we assume that we can generate a headline with an automatic text summarization approach. In the proposed method a headline generator consists of two modules; an encoder and a decoder. First, the encoder constructs intermediate representation from a body of an article. Second, the decoder generates a headline using the intermediate representation. The encoder and the decoder are implemented with neural networks in our proposed method. The method using a neural network need not any linguistic knowledge like existing methods such as part-of-speech tagging, word alignment, and so on. This method is completely data-driven approach.

In Sect. 2 we explain related works including text summarization, headline generation, machine translation and neural language model. In Sect. 3 we explain a neural language model with a recurrent neural network. We especially describe recurrent neural networks and Long Short Term Memory, which are elements of a neural network language model. In Sect. 4 we describe our proposed method constructed with neural networks. The proposed method uses Encoder-Decoder model in machine translation. In Sect. 5 we carry out experiments of headline generation and discuss the results. Finally we remark conclusions and future works in Sect. 6.

2   Related Works

In this section we explain related works, which are text summarization, headline generation and machine translation. We describe only neural network machine translation with respect to machine translation because machine translation is not related to text summarization directly but the proposed method uses neural network machine translation techniques. We describe distributed representation of a text to help you to understand intermediate representation in the proposed method.

2.1 Text Summarization

Text summarization makes an original text to be shorter and more concise text while keeping original semantics. Many approaches estimate an importance of sentence in an article and extract only the most important sentences as an abstract [5, 14, 26]. In this approach, tf*idf [20] is used to calculate importance of a sentence. Other approaches choose sentences according to location where a sentence appears in an article [2, 5]. For example, the first sentence in a paragraph is regarded as an important sentence. Marcu [15] focuses on discourse structure of an article and selects sentences for an abstract. These approaches belong to an extractive approach and their goal is to find a few sentences representing contents of an article.

Rush et al. [19] proposed an abstractive approach using neural network machine translation [11]. The abstractive approach constructs distributed representation from an input text and generates an abstract from the distributed representation. Essentially the approach does not depend on an input text and construct an abstract based on vocabulary in training corpus.

2.2 Headline Generation

The headline expresses the overview of articles with one short sentence. Headline generation aims to make a short sentence from the content of the articles. Banko et al. [3] proposed headline generation task with statistical approach in abstractive way, which understands each text and generates appropriate headline. Recently, some researchers try to generate headlines using neural networks [13], and some of them improve it adding more features. Takase et al. [23] proposed abstractive neural network model considering syntactic and semantic information and Xu et al. [25] uses topic information as additional feature.

2.3 Neural Network Machine Translation

We explain neural network machine translation. Traditional machine translation used statistical machine translation. In statistical machine translation, a phrase table, which denotes matching between phrases in an original language and phrases in a target language, is constructed based on co-occurrence frequency. Sutskever et al. [22] proposed a neural network machine translation model. Their proposed method is called "Encoder-Decoder model". Encoder-Decoder model does not construct the phrase table and make distributed representation of an input sentence. The distributed representation includes all information to translate a sentence. By using the distributed representation, the encoder generates a translated sentence.

2.4 Neural Language Model

We describe a neural network language model, which are basic techniques to realize neural network machine translation. We especially explain a distributed representation, recurrent neural networks, and Long Short Term Memory.

Distributed Representation In distributed representation words and sentences are represented as vectors based on distributed hypothesis [6, 8]. Mikolov et al. [17, 18] proposed word representation in vector space with a neural network language model. The word representation capture semantic similarity of words and realize analogical task with linear algebra of word representation. In Encoder-Decoder model an inputted original sentence is represented as a vector and the vector is used as an initial value in a decoder, which generates a translated sentence, so we can regard the vector as distributed representation of a sentence. However, Encoder-Decoder model cannot deal with long sentences because the length of distributed representation is fixed. Bahdanau et al. [1] proposed an attentional network to overcome the previous problems. They introduced variable length of distributed representation.

Recurrent Neural Networks In natural language processing n-gram model is usually used to construct a language model. Bengio et al. [4] proposed a neural network language model with a multilayer neural network. However, a language model generally has to deal with variable length sentences because sentences consist of the different number of words. Hence, a recurrent neural network [16] which improves the multilayer neural network, is proposed. Encoder-Decoder model and attentional network use Long Short Term Memory (LSTM) [7, 9, 21], which is one of the recurrent neural networks and can deal with variable length sentence well.

3 Recurrent Neural Network Language Model

Recurrent neural networks are used in state-of-the-art natural network language models because the network can deal with variable length of a sentence, especially Long Short Term Memory (LSTM) achieves the highest performance in various natural language processing tasks. We explain simple recurrent neural networks and LSTM.

3.1 Recurrent Neural Network

Recurrent neural networks have a feedback loop on a hidden layer and the input of a hidden layer at time t is the hidden layer at time $t - 1$. An architecture of a recurrent neural network is shown in Fig. 1. The recurrent neural network have a deep architecture as denoted in unfolded form.

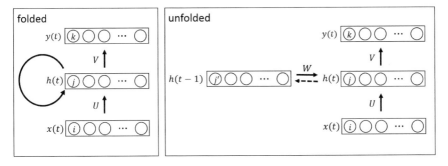

Fig. 1 Recurrent neural networks

$$u_j^t = \sum_i x_i(t)U_{ji} + \sum_{j'} h_{j'}(t-1)W_{jj'} + \text{bias}_{ji} \tag{1}$$

$$h_j(t) = f(u_j^t) \tag{2}$$

$$v_k^t = \sum_j h_j(t)V_{kj} + \text{bias}_{kj} \tag{3}$$

$$y_k(t) = f^{out}(v_k^t) \tag{4}$$

f, f^{out} are activate functions. Because it has a feedback loop, all previous input information is accumulated in the hidden layer.

We show formulas which is trained by backpropagation. E is loss function and δ is a function that differentiated from a loss function. The δ of an output layer unit at time t is

$$\delta_k^{out,t} \equiv \frac{\partial E}{\partial v_k^t} \tag{5}$$

δ of a hidden layer unit at the same time is

$$\delta_j^t \equiv \frac{\partial E}{\partial u_j^t} = \left(\sum_k V_{kj}\delta_k^{out,t} + \sum_{j'} W_{jj'}\delta_{j'}^{t+1} \right) f'(u_j^t) \tag{6}$$

Therefore, gradients for all layers are

$$\frac{\partial E}{\partial U_{ji}} = \sum_{t=1}^{T} \frac{\partial E}{\partial u_j^t}\frac{\partial u_j^t}{\partial U_{ji}} = \sum_{t=1}^{T} \delta_j^t x_i^t \tag{7}$$

$$\frac{\partial E}{\partial W_{jj'}} = \sum_{t=1}^{T} \frac{\partial E}{\partial u_j^t}\frac{\partial u_j^t}{\partial W_{jj'}} = \sum_{t=1}^{T} \delta_j^t h_j^{t-1} \tag{8}$$

$$\frac{\partial E}{\partial V_{ji}} = \sum_{t=1}^{T} \frac{\partial E}{\partial v_j^t} \frac{\partial v_j^t}{\partial V_{kj}} = \sum_{t=1}^{T} \delta_j^t h_i^t \qquad (9)$$

By using these equation update weights with stochastic gradient decent. The α is a learning rate.

$$w^{t+1} = w^t - \alpha \frac{\partial E}{\partial w} \qquad (10)$$

Gradient vanishing and gradient explosion occurs in recurrent neural networks. The gradient vanishing means gradient becomes too small as time goes back. On the other hand, the gradient explosion means gradient becomes too large as time goes back. The gradient vanishing and the gradient explosion make learning to be so difficult. Moreover, when an absolute value of the maximum eigenvalue in feedback loop weights is less than 0, recurrent neural networks are stable but they tend to forget past inputs quickly. It shows that a language model with a recurrent neural network can not deal with a long sentence. These drawback are solved in Long Short Term Memory (LSTM).

Long Short Term Memory (LSTM) LSTM is a neural network that can deal with long distance dependencies and short distance dependencies simultaneously and this is superior to usual recurrent neural networks. The core of LSTM, memory cell, is a very simple recurrent neural network whose feedback loop weight is an identical matrix. Because eigenvalues of the feedback loop weight is equal to 1, past information is kept. Input gate, forget gate, and output gate are added to control the memory cell. Figure 2 shows the LSTM.

In memory cell past information is kept by a feedback path and long distance memory is realized. A range for a gate is from 0 to 1 and when output of a gate is near 1, the gate is opened. When output of a gate is near 0, the gate is closed. The gate execution makes the memory cell to be controlled according to input data. The input gate controls input to memory cell. The forget gate controls past information for memory cell. The output gate controls output from LSTM. Big arrows show inputs to LSTM, which are combined with input from an input layer and a hidden layer in previous step. The following equations show processes of LSTM.

$$h_j^t = g_j^{F,t} h_j^{t-1} + g_j^{I,t} f(u_j^t) \qquad (11)$$

$$u_j^t = \sum_i U_{ji} x_i^t + \sum_{j'} W_{jj'} h_j^{t-1} \qquad (12)$$

$$g_j^{F,t} = f^{gate}(u_j^{F,t}) = f^{gate}\left(\sum_i U_{ji}^F x_i^t + \sum_{j'} W_{jj'}^F g_j^{O,t-1} f(h_j^{t-1}) + h_j^{t-1} \right) (13)$$

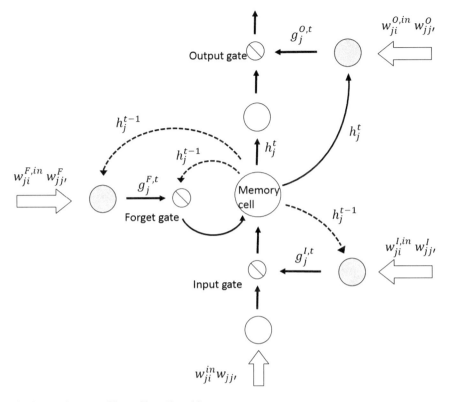

Fig. 2 Architecture of Long Short Term Memory

$$g_j^{I,t} = f^{gate}(u_j^{I,t}) = f^{gate}\left(\sum_i U_{ji}^I x_i^t + \sum_{j'} W_{jj'}^I g_j^{O,t-1} f(h_j^{t-1}) + h_j^{t-1}\right) \quad (14)$$

$$g_j^{O,t} = f^{gate}(u_j^{O,t}) = f^{gate}\left(\sum_i U_{ji}^O x_i^t + \sum_{j'} W_{jj'}^O g_j^{O,t-1} f(h_j^{t-1}) + h_j^t\right) \quad (15)$$

$$f^{gate}(z) = \frac{1}{1 + e^{-z}} \quad (16)$$

The activation function f^{gate} shows sigmoid function to limit the range from 0 to 1. We show training with backpropagation. The total input to output layer is

$$f(x) = tanh(x) \quad (17)$$

$$u_k^{out,t} = \sum_j V_{kj} f(u_j^t) \quad (18)$$

and a gradient of $u_k^{out,t}$ is

$$\frac{\partial u_k^{out,t}}{\partial u^{O,t}} = V_{kj} f'(u_j^{O,t}) f(h_j^t) \tag{19}$$

δ of unit that outputs $g_j^{O,t}$ is

$$\delta_j^{O,t} = f'(u_j^{O,t}) f(h_j^t) \left(\sum_k V_{kj} \delta_k^{out,t} + \sum_{j'} W_{j'j} \delta_{j'}^{t+1} \right) \tag{20}$$

δ of units that act an activation function after the cell is

$$\tilde{\delta}_j^t = g_j^{O,t} f'(h_j^t) \left(\sum_k V_{kj} \delta_k^{out,t} + \sum_{j'} W_{j'j} \delta_{j'}^{t+1} \right) \tag{21}$$

The following equations indicate δ of memory cell unit, the input, the forget gate, and the input gate, respectively.

$$\delta_j^{cell,t} = \tilde{\delta}_j^t + g_j^{F,t+1} \delta_j^{t+1} + \delta_j^{I,t+1} + \delta_j^{F,t+1} + \delta_j^{O,t} \tag{22}$$

$$\delta_j^t = g_j^{I,t} f'(u_j^t) \delta_j^{cell,t} \tag{23}$$

$$\delta_j^{F,t} = f'(u_j^{F,t}) h_j^{t-1} \delta_j^{cell,t} \tag{24}$$

$$\delta_j^{I,t} = f'(u_j^{I,t}) f(u_j^t) \delta_j^{cell,t} \tag{25}$$

After that update weights in LSTM by stochastic gradient descent like recurrent neural networks.

4 Proposed Method

In this section we describe our proposed method, which generates a headline from an article with Encoder-Decoder model [22]. We explain Encoder-Decoder model for headline generation and a headline generation process with the proposed method.

4.1 Encoder-Decoder Model

Encoder-Decoder model was proposed by Sutskever et al. [22] to realize neural network machine translation. In Fig. 3a basic Encoder-Decoder model is shown.

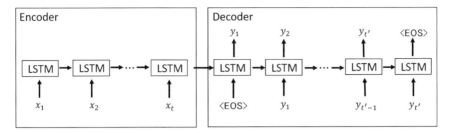

Fig. 3 Text processing with encoder-decoder model

The figure is illustrated with unfolded style of a recurrent neural network. The decoder is implemented with a recurrent neural network. LSTM is expanded with the length of an input sentence. The decoder is implemented like the encoder, so the decoder is expanded with the length of a generated headline. This model is adaptable for various length of input or output because it includes a recurrent architecture.

In this model connection between the encoder and the decoder is sparse and a lot of information is not submitted from the encoder to the decoder. In a bridge between them states in LSTM is submitted to the decoder and the states include semantics of an input sentence. Hence, we regard the states as a distributed representation of a sentence.

$\mathbf{x}$ denotes a sentence in a source language and $\mathbf{y}$ denotes a sentence in a target language in machine translation. $\mathbf{x}$ and $\mathbf{y}$ consist of a sequence of words, $[x_1, x_2, \ldots, x_t]$ and $[y_1, y_2, \ldots, y_{t'}]$ respectively. The word is represented as a vector using weights between an input layer and a hidden layer. The word vector is called distributed representation and the weights are called embedded matrix learned with training data. In a translation step $\mathbf{x}$ is inputted to the encoder and after submission of $\mathbf{x}$, <EOS> is inputted. <EOS> denotes the end of a input sentence and execution of the decoder. Hence, after inputting <EOS>, only the states of the hidden layer in the encoder are sent to the decoder, so a state includes all meanings of the input sentence. The decoder accepts inner states in the decoder and generates a sentence in a target language, a sequence of words. The decoder accepts a predicted word in previous step and predicts the following word. When translation has been finished, the decoder makes <EOS> as a symbol of the end of a sentence. Empirically it is popular to reverse the order of the word of input sentence, so in this paper, we did experiments using forward sentence and reversed sentence.

In training we prepare a corpus which includes many of parallel translation sentences. The model accepts a sentence in a source language and generates a sentence in a target sentences. In this case correct translation denotes $\mathbf{t}$. Using a generated sentence and a correct translation, we define a loss function with cross entropy.

$$E = -\sum_k t_k \log y_k + (1 - t_k) \log(1 - y_k) \tag{26}$$

In backpropagation minimizing Eq. (26), we carry out stochastic gradient descent.

4.2   Headline Generation with Encoder-Decoder Model

We use Encoder-Decoder model for headline generation. The model compresses an input sentence into a fixed-length vector and from the vector an output sentence is generated. Hence, the model is essentially a generative approaches and linkage between input and output is weak, so we use the model to generate a headline from an article. In headline generation input data denotes a body of an article and output data denotes a headline of the article.

5   Experiments

We experiment on a task of headline generation with the proposed method using news articles. In this section, we explain about dataset which we actually used and the detailed explanation of Encoder-Decoder model. Moreover we describe a comparative method, statistical machine translation; MOSES [10]. After that we show the results of these experiments and discussing about that.

5.1   Dataset

We use the dataset called "Thomson Reuters Text Research Collection (TRC2)" which is a collection of documents that appears on Reuters newswire from 2008 to 2009. This data includes about 1,800,000 articles. These news articles include various topics such as politics, sports, governments, and so on. We show some examples of Reuters news articles in Table 1. As for preprocessing we convert a number to "num". In addition, to reduce the size of vocabulary, we convert the word which appeared only once to "UNK" which means unknown word and we convert all the word to the lower-case words. Moreover, this data sometimes includes articles which is written in French, so we do not use such articles that was not written in English in this experiment. Generally, it is good to use the first sentence of the body of the articles, so in this time we also use only first sentence for input. In addition, we remove the articles that has more than 80 words in first sentence. After doing such preprocessing, we select articles that was used for training and testing. In this experiment, we use first 16,959 articles for training and the size of vocabulary is 16,117. As for test data, we use following 7544 articles.

Table 1 Example of bodies of articles

PRESS DIGEST - Philippine newspapers - Jan 11

MANILA, Jan 11 (Reuters) - These are the leading stories in Manila newspapers on Friday. Reuters has not verified these stories. - The Philippine anti-graft court has found 1.1 billion pesos ($27 million) worth of assets in the Jose Velarde bank account of deposed President Joseph Estrada. (ALL PAPERS) - The Armed Forces of the Philippines said it will be forming six new battalions to be trained specifically in fighting the communist New People's Army. (THE PHILIPPINE STAR) - The Senate Committee on Ways and Means ignored objections of the Department of Finance and will go ahead with a hearing seeking the suspension of the 12 percent value added tax on petroleum products. (THE MANILA TIMES, PHILIPPINE DAILY INQUIRER) ******* BUSINESS - The Department of Finance will support measures seeking to increase taxes on alcohol and cigarettes, but is not keen on a proposal to tax text messaging, at least not this year. (PHILIPPINE DAILY INQUIRER) - The Philippines recorded over 3 million tourist arrivals in 2007, with foreign tourists spending a total $4.9 billion in the country, surpassing the contribution of the business process outsourcing sector to the economy. (MANILA STANDARD TODA, BUSINESSWORLD, MALAYA) - The government has earmarked 624.1 billion pesos for debt payments this year, higher than the 612.8 billion pesos that was planned last year, Department of Finance data showed. (MANILA STANDARD TODAY, THE MANILA TIMES, THE PHILIPPINE STAR, PHILIPPINE DAILY INQUIRER) ($1 = 40.74) (Reporting by Karen Lema) ((karen.lema@reuters.com@reuters.net; +63 2 841-8937; Reuters Messaging: http://rosemarie.francisco.reuters.com)) Keywords: PHILIPPINES PRESS/

Foreign brokers place net Japan stock sell orders

TOKYO, Jan 11 (Reuters) - Orders for Japanese stocks placed through 13 foreign securities houses before the start of trade on Friday showed an intention to sell a net 10.3 million shares, market sources said. There were buy orders for 26.3 million shares and sell orders for 36.6 million, they said. (Reporting by Elaine Lies) ((elaine.lies@reuters.com@reuters.net; +81 3 3432 8485; Reuters Messaging: elaine.lies.reuters.com@reuters.net)) Keywords: MARKETS JAPAN STOCKS/ORDERS

Cricket-South Africa 437-4 versus West indies (139) - tea

Jan 11 (Reuters) - South Africa were 437 for four in reply to West Indies first innings of 139 all out at tea on the second day of the third test on Friday. Scores: West Indies 139 (S. Pollock 4-35). South Africa 437-4 (G. Smith 147, J. Kallis 74, A. Prince 73 not out, H. Amla 69). (Reporting by Telford Vice; editing by Patrick Johnston) ((patrick.johnston@reuters.com; +44 207 542 7933; Reuters Messaging: patrick.johnston.reuters.com@reuters.net; + 44 207 542 2604; For the latest Reuters Premier League and international football news see: http://football.uk.reuters.com/)) Please double-click on the newslinks below: For more cricket: [CRIC-LEN] For more South Africa-West Indies cricket: [SPO-CRIC-WINDIES-LEN]

5.2 MOSES

We use MOSES as comparative method. This is often used in a phrase-based statistical machine translation task. To improve the baseline for headline generation task,

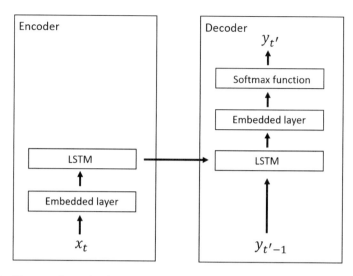

Fig. 4 Architecture of encoder-decoder model

we set an infinite distortion limit. In addition, we cut the generated headline with 10 words which is the average word size of headline length in this dataset.

5.3 Settings

We train the proposed model with backpropagation and we set the number of dimensions as embedded layer = 200 and hidden layer of LSTM = 400. The output layer of decoder is softmax function and outputting one best word. We show a detailed architecture of Encoder-Decoder model that we used in this experiment in Fig. 4. The output layer in decoder is softmax function and it selects best word for every time step.

$$y_k = \frac{\exp(u_k)}{\sum_{j=1}^{k} \exp(u_j)} \tag{27}$$

We implement the proposed method with Chianer [24]. We carry out the proposed method on multicore CPU without GPGPU. The training time for one epoch is about an hour and a half and in this time, we show the results with 10 epochs.

5.4 Results

We carry out experiment using Encoder-Decoder model. We use forward sentence and reversed sentence in encoder. First of all, we show the results with ROUGE that is the metrics about summarization quality. After then we discuss a performance of the proposed method with the example of generated headlines.

Firstly, we show the results measured by ROUGE [12]. ROUGE is often used for measuring summarization quality and we use this measurement for headline generation task to express the quality. It shows the percentage of overlapping words between true headline and generated headline. In this experiment we use ROUGE-1, ROUGE-2, ROUGE-S, and ROUGE-S4. ROUGE-N is represented as the following equation:

$$\text{ROUGE} - \text{N} = \frac{\sum_{\text{gram}_N \in \text{actual headline}} M(\text{gram}_N)}{\sum_{\text{gram}_N \in \text{actual headline}} C(\text{gram}_N)} \tag{28}$$

N indicates the type of N-gram, so ROUGE-1 and ROUGE-2 is unigram and bigram respectively. $M(\text{gram}_N)$ is the number of n-grams matched between true headline and generated headline and $C(\text{gram}_N)$ is the total number of n-grams in true headline. ROUGE-S indicates skip-bigram. Skip-bigram is any pair of words allowing for arbitrary gaps.

$$\text{ROUGE} - \text{S} = \frac{(1 + \beta^2)\text{Rec}_S\text{Prec}_S}{\text{Rec}_S + \beta^2\text{Prec}_S} \tag{29}$$

$$\text{Rec}_S = \frac{\sum_{\text{gram}_S \in \text{actual headline}} M(\text{gram}_S)}{\sum_{\text{gram}_S \in \text{actual headline}} C(\text{gram}_S)} \tag{30}$$

$$\text{Prec}_S = \frac{\sum_{\text{gram}_S \in \text{actual headline}} M(\text{gram}_S)}{\sum_{\text{gram}_S \in \text{generated headline}} C(\text{gram}_S)} \tag{31}$$

β is the harmonic factor between Rec_S and Prec_S. In this experiments we set very large number as β. Similarly, $M(\text{gram}_S)$ is the number of skip-bigram matched between actual headline and generated headline and $C(\text{gram}_S)$ is the total number of skip-bigram. ROUGE-S4 is different from ROUGE-S which consider skip-bigram only within 4 words.

Table 2 shows the results with ROUGE. As shown in Table 2, we can get better results with Encoder-Decoder model than using MOSES with all metrics. Especially among them, we can get 2.2 points higher in ROUGE-1, and about 1 points higher in other metrics when using reversed sentence for input than using forward sentence.

Table 2 Evaluation of generated headlines with ROUGE

	ROUGE-1	ROUGE-2	ROUGE-S	ROUGE-S4
MOSES	23.5	10.0	7.6	8.0
Forward sentence	32.9	23.5	21.7	21.6
Reversed sentence	**35.1**	**24.6**	**22.4**	**22.3**

Table 3 Examples of generated headlines

Actual headline	Forward sentence	Reversed sentence	MOSES
Foreign brokers place net Japan stock sell orders	Foreign brokers place net Japan stock sell orders	Foreign brokers place net Japan stock sell orders	Update num Japan foreign securities houses before start to selll
PRESS DIGEST - Philippine newspapers - Jan 11	Press digest - Philippine newspapers - Jan num	Press digest - Philippine newspapers - Jan num	Philippine num press digest - Manila newspapers - on
US STOCKS-Wall St opens lower on bank jitters	UNK stock market is a UNK UNK on a UNK	Us stocks-wall st opens flat on economic worry, recession	Source document can be at http://www.sebi.gov.in dp
Cricket-South Africa 437-4 versus West indies (139) - tea	Cricket-India to continue Australia tour	Cricket-South Africa num versus West Indies (num) - close	Update num rwe might rweg.de priority - sixth partner in
Eight dead in Mexico helicopter crash	UNK leading to Georgia's election	Mexican retailers to cut prices on num UNK	Update num Tennis mum Lindsay Davenport's new disciple
Colorado governor outlines agenda, skips cost	No agreement at Western Sahara EUR	UNK hurt as UNK ipo	By Keith Coffman Denver num colorado gov.

Next, we see the performance of generated headlines in more detail. Table 3 shows the examples of results which was generated by Encoder-Decoder model and MOSES.

Overall, we can get better results when we use Encoder-Decoder model compared when we use MOSES. We get a lot better results with reversed sentences as input than using forward sentences.

From now, we will see the results with a focus on the headlines which was generated by Encoder-Decoder model. The first and the second line in Table 3 can generate a correct headline of an article. The third and the fourth lines show reversed version can predict some correct words but the normal version cannot predict any words. Finally, regarding the last two lines they generate are completely different from the actual headline in both cases. When we check all generated headlines, we can see any results which is able to generate better results with reversed sentence. In addition, it is rare to find bad results only in the reversed version, so we can find it is effective to use reversed sentences in this study.

6 Conclusion

In this study we make a headline from a body of an article with Encoder-Decoder model, which was proposed in machine translation. We achieve good headline generation in some articles and also get bad headline generation in other articles. We think the results depends on similarity between training corpus and test sentences, so we have to discuss various training data configuration. However, ROUGE score reveals proposed approach is more effective than MOSES. In addition, using reversed sentence is also somewhat effective especially for metrics of ROUGE-1 with 35.1 points than using forward sentence. In future work we will try to improve the accuracy of headline generation task. In this time, we use only first sentence for input, but we want to try using full sentence information to generate headlines. We will use attention mechanism and try to generate better headlines. In addition, this news data contains various topics like sports, politics and so on, so in next time we would like to utilize these information effectively.

References

1. Bahdanau, D., Cho, K., Bengio, Y.: Neural machine translation by jointly learning to align and translation. In: Proceedings of ICLR2015 (2015)
2. Brandow, R., Mitze, K., Rau, L.: Automatic condensation of electronic publications by sentence selection. Inf. Process. Manag. **31**(5), 675–685 (1995)
3. Banko, M., Mittal, V.O., Witbrock, M.J.: Headline generation based on statistical translation. In: Proceedings of ACL00, pp. 318–325 (2000)
4. Bengio, Y., Ducharme, R., Vincent, P., Jauvin, C.: A neural probabilistic language model. J. Mach. Learn. Res. **3**, 1137 (2003)
5. Edmundson, H.: New methods in automatic abstracting. J. ACM **16**(2), 264–285 (1969)
6. Firth, J.: A synopsis of linguistic theory 1930-1955. Stud. Linguist. Anal. pp. 1–32 (1957)
7. Graves, A., Schmidhuber, J., Cummins, F.: Learning to forget: continual prediction with LSTM. Neural Comput. **12**(10), 2451–2471 (2000)
8. Harris, Z.: Distributed structure. Word **10**(23), 146–162 (1954)
9. Hochreiter, S., Schmidhuber, J.: Long short-term memory. Neural Comput. **9**(8), 1735–1789 (1997)
10. Koehn, P., Hoang, H., Birch, A., Callison-Burch, C., Federico, M., Bertoldi, N., Cowan, B., Shen, W., Moran, C., Zens, R.: Moses: open source toolkit for statistical machine translation. In: Proceedings of ACL, pp. 177–180 (2007)
11. Koehn, P.: Statistical Machine Translation. Cambridge University Press, Cambridge (2009)
12. Lin, C.: Rouge: a package for automatic evaluation of summaries. In: Proceedings of ACL, pp. 74–81 (2004)
13. Lopyrev, K: Generating news headlines with recurrent neural networks. arXiv preprint arXiv:1512.01712v1 (2015)

14. Luhn, H.: The automatic creation of literature abstracts. IBM J. Res. Dev. **2**(2), 159–165 (1958)
15. Marcu, D.: From discourse structures to text summaries. In: Proceeding of the ACL Workshop on Intelligent Scalable, Text Summariztion pp. 82–88 (1997)
16. Mikolov, T., Karafiat, M., Burget, L., Cernocky, J.H., Khudanpur, S.: Recurrent neural network based language model. In: Proceedings of INTERSPEECH2010 (2010)
17. Mikolov, T., Yih, W., Zweig, G: Linguistic regularities in continuous space word representations. In: Proceedings of HLT-NAACL (2013)
18. Mikolov, T., Chen, K., Corrado, G., Dean, J.: Efficient estimation of word representation in vector space. In: Proceedings CoRR (2013)
19. Rush, A.M., Chopra, S., Weston, J.: A neural attention model for sentence summarization. In: Proceedings of EMNLP 2015 (2015)
20. Salton, G.: Automatic Text Processing. Addison-Wesley, Boston (1989)
21. Sundermeyer, M., Schluter, R., Ney, H.: LSTM Neural Networks for Language Modeling INTERSPEECH (2010)
22. Sutskever, I., Vinyals, O., Le, V.Q.: Sequence to sequence learning with neural network. In: Proceedings of NIPS2014 (2014)
23. Takase, S., Suzuki, J., Okazaki, N., Hirao, T., Nagata, M.: Neural headline generation on abstract meaning representation. In: Proceedings of EMNLP, pp. 1054–1059 (2016)
24. Tokui, S., Oono, K., Hido, S., Clayton, J.: Chainer: a next-generation open source framework for deep learning. In: Proceedings of NIPS2015 (2015)
25. Xu, L., Wang, Z., Ayana, Liu, Z., Sun, M.: Topic sensitive healine generation. arXiv preprint arXiv:1608.05777v1 (2016)
26. Zechner, K.: Fast generation of abstracts from general domain text corpora by extracting relevant sentences. In: Proceedings of the 16th International Conference on Computational Linguistics, pp. 986–989 (1996)

Customer State Analysis with Enthusiasm Analysis

Hidekazu Yanagimoto

Abstract In this paper I propose customer behavior analysis using enthusiasm analysis, which estimates customers' activation levels. Finally our goal is to discovery drop-off users from access logs according to the activation level. It is important to find drop-off users earlier and pay attention to them to stay in a service. Usually prediction models are constructed with machine learning and RFM analysis is used in marketing field. In this paper I estimate enthusiasm levels, which denote customers' activation, from observations and apply them to prediction of discovery of drop-off users. In evaluational experiments I use real online shop access logs and discuss relation between enthusiasm levels and drop-off users. I confirmed that many drop-off users took lower enthusiasm levels in evaluation point and the enthusiasm level could be used to predict drop-off users.

Keywords Customer relation management · Access log analysis · Probabilistic modeling

1 Introduction

Theses days there are many kinds of online shops in the Internet and the shops compete each other to get larger market because shops deal with the same products. In this case it is important to obtain new customers, which will buy products in future, by service registration and to make customers to stay in the online ships. To achieve the aims the shops make their services to satisfy them. Hence, in business management many researchers pay attention to customer satisfaction and there are many researches on customer relation management [1–3]. However, the approaches is to understand customer behavior by data analysis methods and it is difficult to apply them to prediction customers' drop-off directly.

H. Yanagimoto (✉)
College of Sustainable Systems Science, Osaka Prefecture University,
1-1, Gakuen-cho, Naka-ku, Sakai, Osaka 599-8531, Japan
e-mail: hidekazu@kis.osakafu-u.ac.jp

© Springer International Publishing AG 2018
T. Matsuo et al. (eds.), *New Trends in E-service and Smart Computing*,
Studies in Computational Intelligence 742, https://doi.org/10.1007/978-3-319-70636-8_7

Big Data trends cause data analysis of logs online shops store as histories of users in information science. In the approach it is a main topic for the online shops to extract information on users from the logs. For example, the logs consist of access logs and purchase histories. Many recommendation systems are developed with the purchase histories. It is called basket analysis. One of results of log analysis to predict customer behavior is a royal customer discovery task. Online shops want to find royal customers, who are customers buying many products in the shop, and promote other products for them. The royal customers visit online shops frequently and spend much money to buy products in the shop. So online shops recommend products to the royal customers because of increasing their sales and make them to keep staying at the online shops by giving them better supports. The approach is one of customer segmentations and customer targeting and many researches deal with the topic. On the other hand, the online shops must keep a community of users registered in the shop because they keep their market. To achieve the aim the shop have to discover drop-off users as possible as soon. But there are a few researches dealing with drop-off customer discovery.

To predict future customers' behaviors usual approaches use machine learning, Linear Regression [4], Support Vector Machine [5], Random Forest [6], and Gradient Boosting Decision Tree [7]. These approaches construct a nonlinear function from training data. In Linear Regression classifier designers can discuss which features are effective to predict labels. But in other methods it is difficult to discuss constructed models because the models are too complex. Moreover, all approaches needs training data with correct labels and making the training needs many efforts.

In this paper I discuss customer behavior analysis with enthusiasm analysis [8] to discovery of drop-off customers from their access logs. I developed enthusiasm analysis, which estimate user's activation level from observations without correct labels assuming a Poisson distribution model. In enthusiasm analysis you can assign active state or inactive state to observation. Focusing on enthusiasm levels of users, I discuss relation between the enthusiasm level and customer's drop-off. Using access logs in an online shop I carried out experiments to evaluate enthusiasm analysis from the viewpoint of relation between the enthusiasm level and drop-off. I confirmed that enthusiasm analysis could predict drop-off users.

In Sect. 2 I describe related works on customer relation management and burst analysis. In Sect. 3 I explain burst analysis, which is a basic method of enthusiasm analysis. In Sect. 4 I propose enthusiasm analysis. In Sect. 5 I carry out experiments and discuss a performance of the proposed method. In Sect. 6 I conclude this paper and describe future works.

2   Related Works

In this section I describe usual customer analysis method, RMF analysis and burst analysis, which estimate active states of events from observation with a probabilistic model.

Fig. 1 An example of RFM
analysis

Frequency / Recency	More than 20	More than 10	More than 1
Within 1 week	Royal Customers		
Within 1 month			
Within 2 months			Drop-off Customers

To find royal customers RFM analysis [9], which is an analysis method based on three measures, recency, frequency, and monetary, is often used in marketing research because the measures are important to classify customers. In RMF analysis the three measures are counted from access logs. Hence, the measures depends on observations strongly and there is not a prediction approach to find drop-off users because the method is an analysis method but not a predictor. Moreover, RMF analysis is sensitive to a threshold and it is so difficult to define the appropriate threshold manually because the threshold affects a performance of RFM analysis. Figure 1 shows an example of RFM analysis using frequency and recency.

Machine learning approaches construct a predictor from training data with correct labels. Linear regression and Support Vector Machine (SVM) use a linear discriminant function. Random Forest and Gradient Boosting Decision Tree use many decision trees to construct a nonlinear discriminant function. Off course SVM can constructs a nonlinear discriminant function using kernel functions. These approaches can achieve high prediction precision but constructing training data with correct labels need many efforts.

Another approach is a method to construct a latent structure from access logs with a data generation model and extract knowledge from the structure. One of user's state prediction methods is burst analysis [10]. In burst analysis event occurs under a probabilistic distribution. Especially burst analysis assumes that event occurrence interval is determined under exponential distributions. Hence, essentially burst analysis find allocation of the exponential distributions form observations using a dynamic programming algorithm [11]. Izumi et al. [12] proposed query analysis based on query submission frequency. The query analysis method used burst analysis and they confirmed query submission was increased according to real events. Yanagimoto et al. proposed enthusiasm analysis [8], which is improvement of burst analysis estimating inactive states and active states simultaneously. In this paper I use enthusiasm analysis to predict customer behavior.

3 Burst Analysis

I explain burst analysis proposed by Kleinberg because burst analysis is a basic idea of enthusiasm analysis. Because in original paper burst analysis is applied to e-mails, I explain burst analysis using e-mail analysis.

Burst analysis assumes message arrival interval depends on an exponential distribution $f_i(x)$.

$$f_i(x) = \alpha_i e^{-\alpha_i x} \qquad \alpha_i > 0 \tag{1}$$

In Eq. (1) i denotes a burst level, which shows how actively messages are generated. So i is related to an average frequency of message generation. x denotes message generation interval observed from real e-mail arrival. The exponential distribution is high in $x = 0$ and as α_i is small, a probability of x except 0 is high. When many messages are observed, the messages are generated from the exponential distribution with a large α_i. Hence, a problem is how you predict each message is generated with the distribution from observed data.

Burst analysis uses maximum likelihood estimation to predict a model generating observations. Previously we define α_0 under the assumption of completely uniform message arrival during observations. Hence, we define α_i below.

$$\alpha_0 = \left(\frac{n}{T}\right)^{-1} \tag{2}$$

$$\alpha_i = \alpha_0 s^i \qquad s > 1.0 \tag{3}$$

In Eq. (2) n denotes the number of total arrival messages and T denotes a period of all message arrival. Hence, $\frac{n}{T}$ shows the average number of arrival messages. Since α_i is large as i is large, $f_i(x)$ is a distribution generating messages with a short interval. The maximum state is defined according to the minimum interval in observations. So it is calculated with Eq. (4).

$$k = \lceil 1 + \log_s T + \log_s \delta(X)^{-1} \rceil \tag{4}$$

$\delta(X)$ means the minimal intervals in observations. Using Eq. (4) we can set the maximum number of states in observed data.

Figure 2 shows a burst level estimation model for all observations in Burst analysis. States in Fig. 2 mean burst levels and each state corresponds to α_i in the exponential distribution $f_i(x)$. In burst analysis we determine one of states in each message as the distribution describes the message generation interval as correctly as possible. So a sequence of the selected states represents burst level transition.

The burst analysis introduces transition weight in the burst model. If there is no transition weight, transition of states for observations is too violet and we cannot capture a latent structure of them. The transition weight $\tau(i, j)$ is defined depending on transition of states below.

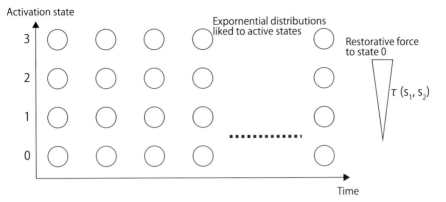

Fig. 2 Burst level estimation model

$$\tau(i, j) = \begin{cases} 0 & (i \geq j) \\ (j - i)\gamma \ln n & (i < j) \end{cases} \tag{5}$$

γ is a predefined parameter and have to satisfy $\gamma > 0$. The transition weight gives transitions moving to a higher state a penalty as rapid transitions do not happen. On the other hand transitions moving to a lower state have no penalty and happens easily.

Finally we find an optimal transition to minimize the following cost function over the observed period.

$$c(\mathbf{q}|\mathbf{x}) = \sum_{t=0}^{n-1} \tau(i_t, i_{t+1}) + \sum_{t=1}^{n} \ln f_{i_t}(x_t) \tag{6}$$

$\mathbf{q}$ denotes a state sequence $(i_1, i_2, \ldots, i_n)$ and $\mathbf{x}$ means a sequence of observed messages arrival intervals $(x_1, x_2, \ldots, x_n)$. Hence, the burst analysis is to find a optimal sequence that minimizes Eq. (6) using observed messages.

We achieve the optimization of Eq. (6) with Viterbi algorithm [11]. The Viterbi algorithm is one of dynamic programing methods and consists of a forwarding step and a backward step. In the forwarding step the algorithm calculates costs of all states because a cost of a focusing state depends on costs in the previous message. The cost function is rewrite like Eq. (7).

$$c(\mathbf{q}|\mathbf{x}) = c(\mathbf{q}_{1,\ldots,n-1}|\mathbf{x}_{1,\ldots,n-1}) + \tau(i_{n-1}, q_n) + \ln f_{q_n}(x_n) \tag{7}$$

Using Eq. (7) you can calculate a cost for each state using costs of previous states. This step shows concrete process of a forwarding step in Viterbi algorithm.

In the backward step using the costs calculated in the forwarding step, the algorithm searches the optimal path to minimize Eq. (6). Since the forward step calculates all costs of all states, you can find the optimal cost function. Using Eq. (7) you extract the optimal path to achieve the minimal cost function. In Fig. 2 shaded circles mean selected stated from observed messages using the Viterbi algorithm.

4 Proposed Method

In this section I explain Poisson distribution based active state estimation which improve burst analysis to estimate both active states and inactive states. The improved points are (1) constructing an event occurrence model using Poisson distributions, (2) adding some states corresponding to states below average occurrence, and (3) defining a cost function regarding inactive states. Through the improvements the proposed method can discuss more detailed inactive states that burst analysis collapsed in burst level 0.

4.1 Poisson Distribution Based Active State Estimation

Burst analysis [10] focuses on event occurrence interval time and constructs an event occurrence model with exponential distributions. The analysis, however, has some shortcomings: (1) burst level 0 includes various user's states, (2) duration between the last event occurrence and evaluation is neglected because interval cannot be calculated. To remedy these shortcomings we proposed Poisson distribution based activation state estimation.

I construct an event occurrence model with Poisson distributions instead of exponential distributions because the model can use information after the last event occurrence. Because the proposed model focuses on occurrence frequencies per a predefined interval, it can deal with information after the last event as no events directly. A Poisson distribution is defined below.

$$p(k|\lambda) = \frac{\lambda^k}{k!}e^{-\lambda} \tag{8}$$

Since in the model I regard observations after the last visit as $k = 0$ in the Poisson distribution, the new model can deal with information after the last visit naturally. Figure 3 shows the Poisson model. On the other hand, using exponential distributions like burst analysis, I have to define an interval after the last visit. When the interval is duration between the last visit and evaluation, the interval is often underestimated. When the interval is enough long, the interval tends to be overestimated. Figure 4 shows the above problem. So the interval estimation is very a difficult problem and you need more information on users than observations to estimate the interval appropriately. The proposed method is superior to burst analysis to overcome the

Fig. 3 A model with
Poisson distribution

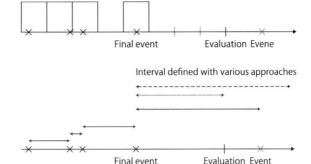

Final event   Evaluation Evene

Fig. 4 Illustrate relation
between evaluation and
interval defined with various
approaches

Interval defined with various approaches

Final event   Evaluation Event

problems. Moreover, in the model I add states corresponding to occurring events infrequently to analyze inactive users.

A model of the proposed method is showed in Fig. 5. The circles in Fig. 5 mean activation states represented with Poisson distributions having different parameter, λ_i. The λ_i is defined below.

1. λ_0 is average occurrence frequency during whole observation.

2. λ_{-2} is lower bound and a user previously defines it considering situation where a system judges drop-off users.
3. λ_i $(i \neq 0, -2)$ is defined below.

$$\lambda_i = \lambda_0 \alpha^i \quad (\alpha > 1) \tag{9}$$

I assume an optimal transition of activation states minimizes the following cost function from observations. The cost function considers likelihood of Poisson distribution assignment and a restorative term.

$$F(s_1, \ldots, s_N | x_1, \ldots, x_N) = \sum_{i=1}^{N} \ln p(x_i | \lambda_{s_i})$$
$$+ \sum_{j=1}^{N-1} \tau'(s_j, s_{j+1}) \tag{10}$$

$(s_1, \ldots, s_N)$ means series of activation states from the first observation to the last observation, $(x_1, \ldots, x_N)$ and $s_i \in \{-2, -1, 0, 1, \ldots\}$. $\tau'(\cdot, \cdot)$ denotes a activation state transition cost that is proportional to a gap between two activation states. The $\tau'(\cdot, \cdot)$ especially makes an activation state go to zero. $\tau'(s_i, s_{i+1})$ is defined below.

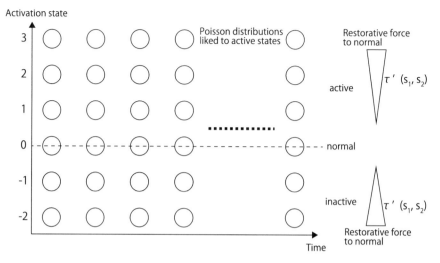

Fig. 5 Active state estimation model in the proposed method

$$\tau'(s_i, s_{i+1}) = \begin{cases} \gamma|s_{i+1} - s_i| \ln N & (s_{i+1} \geq s_i \geq 0, \\ & 0 \geq s_i \geq s_{i+1}) \\ \gamma|s_{i+1}| \ln N & (s_{i+1} \geq 0 \geq s_i, \\ & s_i \geq 0 \geq s_{i+1}) \\ 0 & \text{(otherwise)} \end{cases} \tag{11}$$

The optimization of Eq. (10) is executed with dynamical programming because the cost is rewritten below.

$$F(s_1, \ldots, s_i | x_1, \ldots, x_i) = \ln p(x_i | \lambda_{s_i}) + \tau'(s_{i-1}, s_i) \tag{12}$$
$$+ F(s_1, \ldots, s_{i-1} | x_1, \ldots, x_{i-1})$$

Enthusiasm analysis needs only observations and does not need labels for training data. Hence, enthusiasm analysis can be applied to corpus without any efforts. So it is a strong point comparing with usual machine learning approaches.

5 Experiments

In this section I carry out some evaluation experiments that estimate enthusiasm levels of users in shops with customer purchase histories. I focus on user's frequency of purchases in a shop and estimate the enthusiasm levels. I confirmed that the proposed method could estimate appropriate enthusiasm levels and the estimated levels were related to users' behavior in future.

5.1 Dataset

In experiments I used purchase histories gathered from 2013/1/6 to 2015/6/30 in a real shop. The purchase histories include about 3 million records made by about 23,000 customers in the shop. The purchase histories are regarded as activities histories in the shop. I regard users who do not buy any products from 2015/1/1 to 2015/6/30 as drop-off customers. And I discuss whether the proposed method can predict drop-off customers or not. Table 1 shows details of the purchase histories. In the dataset about 20% of all customers are regarded as drop-off customers. Of course, some of the drop-off customers might buy some products in future but in the experiments I do not consider it at all.

5.2 Parameter Settings

Some parameters are included in enthusiasm analysis. Parameter setting in the experiments is showed in Table 2. In the experiments I set a week to construct an input because I assume that user's behavior depends on a week. I assume a drop-off customer candidates do not by any products in the shop during half of a year. So I defined λ for the lowest state as $\frac{1}{30}$ manually. And the lowest state is defined as enthusiasm level -2. Hence, $\lambda_{-2} = \frac{1}{30}$. The enthusiasm level of customers are determined according to whether they buy any products in the shop or not and the proposed method does not use any other information, purchase price and demographical information of the customers.

The highest enthusiasm level in the proposed method is calculated using observations and parameters below.

Table 1 Details of purchase histories

Customers	23,457
Drop-off customers	5,050

Table 2 Parameters in enthusiasm analysis

Parameter	The proposed method
α	2
γ	1.0
Interval to construct an input	A week
The lowest state	-2
λ for the lowest state	$\frac{1}{30}$

$$i_{max} = \left\lceil \log_2 \frac{\max(x_i)}{\lambda_0} \right\rceil \tag{13}$$

Hence, the number of estimating enthusiasm levels is $i_{max} + 2$.

5.3 Results

Table 3 shows estimated enthusiasm levels of customers and distribution of drop-off customers. Focusing only customers with negative enthusiasm levels, the proposed method detects about 60.9% of all drop-off customers and detection precision achieves about 78.5%. Hence, enthusiasm analysis can predict drop-off customers with high precision. However, about 40% drop-off customers stay in enthusiasm level 0 and it is difficult to detect the drop-off customers with the proposed method. To improve the performance I have to discuss parameters of enthusiasm analysis. For example, more enthusiasm levels are used to analyze the purchase histories. But it is one of future works.

I pick up some characteristic customers and discuss relation between their enthusiasm levels and their purchase activities from Figs. 6, 7, 8, 9 and 10. Especially I discuss enthusiasm level transition patterns. In every figures box graphs denote purchase frequency in a week and cross marks denote enthusiasm level. If I can find some relations between them, more information on drop-off customers can be extracted from enthusiasm analysis.

Figure 6 shows enthusiasm level transition for a customer which buy some products in a shop constantly. The customer is not a drop-off customer in this experiment. The customer constantly visits a shop and buys some products from 2013/1 to 2014/11. So the enthusiasm level is equal to an average level 0.

Figure 7 shows enthusiasm level transition for a drop-off customer. The customer frequently visited a shop at first but does not visit it at the end of observation. Since 2014/4 the customer have not visited a shop. Hence, the enthusiasm level is 1 at the early period and is −2 at the later period. Focusing on the enthusiasm level, I can find customer behavior change form enthusiasm level transition. So you think what

Table 3 Drop-off customers prediction using enthusiasm level

Enthusiasm level	Customers	Drop-off customers
−2	1544	1381
−1	2375	1695
0	19,256	1971
1	264	3
Total	23,440	5050

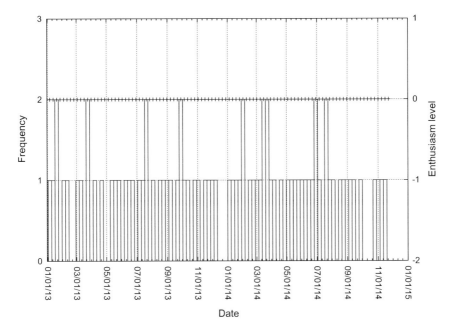

Fig. 6 Transition of enthusiasm level for a non drop-off customer

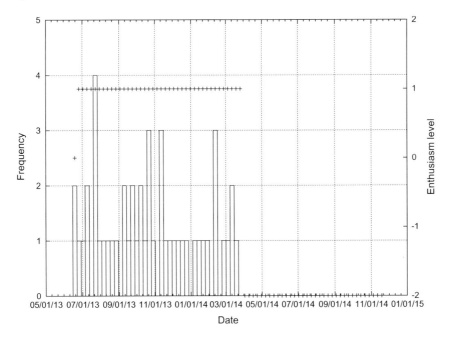

Fig. 7 Transition of enthusiasm level for a drop-off customer

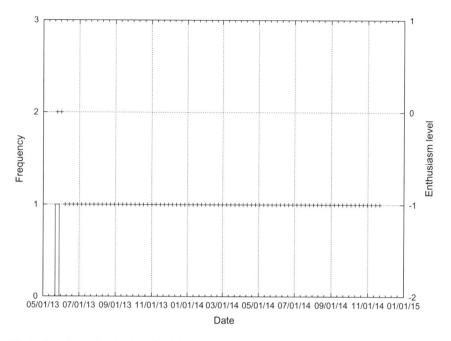

Fig. 8 Transition of enthusiasm level for a drop-off customer who visits a shop once

happens for the customer to avoid a shop. If you can find such customers' behavior change, you can approach them not to leave a shop.

Figure 8 shows enthusiasm level transition for a drop-off customer. The customer's behavior is more extreme than the previous drop-off customer. The customer visits a shop once during observation. Hence, the enthusiasm level keep -1. Because in this case an average of visit frequency is very low, λ_0 is small and enthusiasm analysis does not predict enthusiasm level -2.

Figure 9 shows enthusiasm level transition for a customer who visits a shop again after not visiting a shop for a while. Some customers visit a shop for some months during observation but visit a shop again. The enthusiasm level transition pattern shows that though some customers achieve a low enthusiasm level, they get higher enthusiasm level after some months. For the customers enthusiasm level is not good criterion to predict drop-off customers. Hence, I have to discuss reasons why the enthusiasm level reflect customers behavior. Hence, I discuss other information on a customer, demographical data, purchase price, and so on.

Figure 10 shows enthusiasm level transition for another drop-off customer. It is difficult to predict whether the customer is a drop-off customer or not because the customer visits a shop constantly. To analyze such the customer I have to use other information like the previous customer.

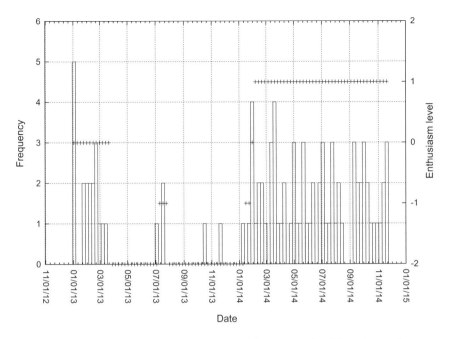

Fig. 9 Transition of enthusiasm level for a non drop-off customer who visits a shop again

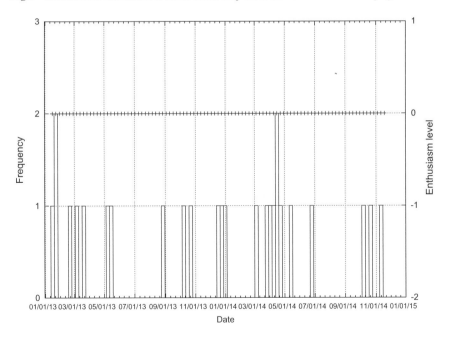

Fig. 10 Transition of enthusiasm level (irregular user)

6 Conclusions

In this paper I applied enthusiasm analysis to customer behavior analysis, which predicted whether a user would be a drop-off user or not. Usually enthusiasm analysis is applied to data analysis and have not been applied to prediction. I carried out some experiments to discuss relation between enthusiasm levels and drop-off and confirmed that enthusiasm level was related to customer behavior prediction strongly. So precision of drop-off customer prediction achieves about 78.5% and coverage is about 60.9%. However, it is difficult to predict some customer behavior prediction from their enthusiasm levels because some customers dynamically move between a high enthusiasm level and a low enthusiasm level. For such the customers I needs to use other customer information, for example, customers' demographical information, purchase histories, and so on.

In future work, I will discuss enthusiasm level transition and classify customers according to their transition pattens. Moreover, I will combine enthusiasm analysis with other machine learning, Gradient Boosting Decision Tree, Deep Learning to achieve higher precision of drop-off customer prediction.

References

1. Fornell, C., Johnson, M.D., Anderson, E.W., Bryant, B.E.: The American customer satisfaction index: nature purpose, and findings. J. Mark. **60**, 7–18 (1996)
2. Flanagan, C.J.: The critical incident technique. Psychol. Bull. **51**, 327–358 (1954)
3. Roos, I.: Method of investigating critical incidents: a comparative review. J. Serv. Res. **4**(3), 192–204 (2002)
4. Hastie, T., Tibshirani, R., Friedman, J.: The Elements of Statistical Learning. Springer, Berlin (2001)
5. Cortes, C., Vapnik, V.: Support-vector networks. Mach. Learn. **20**(3), 273–297 (1995)
6. Ho, T.K.: Random decision forests. In: Proceedings of the 3rd International Conference on Document Analysis and Recognition, pp. 278–282 (1995)
7. Friedman, J.H.: Greedy function approximation: a gradient bossting machine. Ann. Statist. **29**(5), 1189–1232 (2001)
8. Yanagimoto, H.: Customer state estimation with Poisson distribution model. In: Proceedings of AROB2015 (2016)
9. Buckinxa, W., Van den Poel, D.: Customer base analysis: partial defection of behaviourablly loyal lients in a non-contractual FMCG retail setting. Eur. J. Oper. Res. **164**, 252–268 (2005)
10. Kleinberg, J.: Bursty and hierarchical structures in streams. In: Proceedings of the Eighth ACM SIGKDD International Conference on Knowledge Discovery and Data Mining, pp. 91–101 (2002)
11. Viterbi, A.J.: Error bounds for convolutional codes and an asymptotically optimum decoding algorithm. IEEE Trans. Inf. Theory **13**(2), 260–269 (1967)
12. Izumi, A., Yanagimoto, H., Yoshioka, M.: Query analysis for site visits with burst analysis. In: Proceedings of ICKM 2015, pp. 384–389 (2015)